PLANTS — organisms with

Gymnosperms
(conifers and
evergreens)

Pteridophytes
(ferns and horsetails)

Angiosperms
(flowering plants)

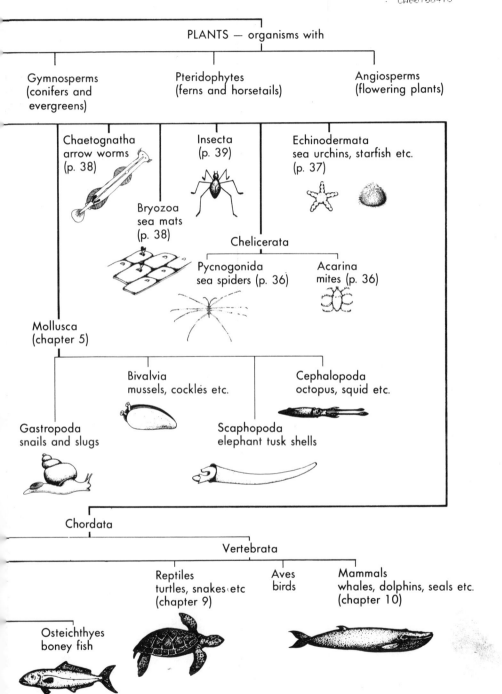

Chaetognatha
arrow worms
(p. 38)

Insecta
(p. 39)

Echinodermata
sea urchins, starfish etc.
(p. 37)

Bryozoa
sea mats
(p. 38)

Chelicerata

Pycnogonida
sea spiders (p. 36)

Acarina
mites (p. 36)

Mollusca
(chapter 5)

Bivalvia
mussels, cocklés etc.

Cephalopoda
octopus, squid etc.

Gastropoda
snails and slugs

Scaphopoda
elephant tusk shells

Chordata

Vertebrata

Reptiles
turtles, snakes etc
(chapter 9)

Aves
birds

Mammals
whales, dolphins, seals etc.
(chapter 10)

Osteichthyes
boney fish

The Seafarer's Guide to
Marine Life

The Seafarer's Guide to
Marine Life

Paul V. Horsman

CROOM HELM
London & Sydney

© 1985 Paul V. Horsman
Croom Helm Ltd, Provident House, Burrell Row,
Beckenham, Kent BR3 1AT
Croom Helm Australia Pty Ltd, First Floor, 139 King Street,
Sydney, NSW 2001, Australia
Distributed in New Zealand by Benton Ross Publishers Ltd,
PO Box 33-055, Takapuna, Auckland 9, New Zealand

British Library Cataloguing in Publication Data

Horsman, Paul V.
    The seafarer's guide to marine life.
    1. Marine biology
    I. Title
    574.92   QH91

ISBN 0-7099-3715-6

FOR CAROL.

Typeset by Words & Pictures Ltd., London SE19.
Printed and bound in Great Britain

Contents

List of Maps

List of Figures and Tables

Figures

Tables

List of Black and White Plates

List of Colour Plates

Foreword

The Director of The Marine Society

A highly qualified reader of this book when it was in manuscript commented, 'I like this book. Some of the descriptions of the animals are exceptionally clear, bearing witness to direct observations. Indeed, the author is at his best with an exciting, first-hand account of bioluminescence, for example. Parts have the genuine excitement of a living early nineteenth-century explorer, a novel occurrence these days.'

This is fair comment. At first sight the book may seem little more than a useful directory. This it is, of course, and it is intended for use by both amateurs and professionals as a means of identifying the plants and animals in the sea. If it were no more than this, it would be well worth publishing, since no other such practical and comprehensive guide exists. But in addition to being a well qualified marine biologist, who has had much experience round the coast, Paul Horsman has been at sea almost continuously for five years in a variety of merchant ships, operating in many different parts of the world. He has therefore had an opportunity, unique in these times, to observe plants and animals at sea, whether seen from the deck, observed while skin-diving in various ports, obtained by way of the fire hydrant or in consequence of netting, or removed from a ship's coolers, condensers, sea water intakes or filters.

Between voyages Paul Horsman has also made a close study of the reports on marine life submitted to the Meteorological Office by seafarers over a period of more than one hundred years — the first such study ever made. These reports have been of great help to him in the construction of his world distribution maps.

There is much in these pages to fascinate any reader, whether it be the parasitic barnacle which castrates a crab; the 'right-handed' and 'left-handed' *Physalia*, which have evolved to travel clockwise or anti-clockwise according to which side of the equator they find themselves on; the electric ray which can develop an electric charge as high as 220 volts; or the tiger sharks whose stomachs have yielded cans, bottles, clothes, shoes, half a crocodile and even undetonated explosives. Paul Horsman moves through this unfamiliar world with ease, wearing his learning lightly.

The Marine Society, the world's oldest maritime charity, has long employed tutors at sea to teach on board merchant ships. Paul Horsman was recommended in this capacity by Dr Frank Evans of the Dove Marine Laboratory, himself at one time a Merchant Navy officer, and his employment was made possible by financial help received from the Leverhulme Trust. It was

The Marine Society which felt that this book was needed and encouraged Paul Horsman to write it. The book also owes much to the artwork of David Henderson and Stephen Devane, who was employed by The Marine Society when the book was being written, and to the drawings of Peter Baker, a professional seafarer at sea with P & O and also very much in touch with The Marine Society.

Ronald Hope

Acknowledgements

I am grateful to The Marine Society and in particular to Dr Ronald Hope for support, encouragement and giving me time to write this book. I thank Dr Frank Evans of the Dove Marine Laboratory, University of Newcastle upon Tyne, for introducing me to The Marine Society and for his help in my research. I am also grateful to Mr Denis McBrearty of the Department of Anatomy at Cambridge University for supplying all the information for Chapter 10, and for his help and useful criticism.

My thanks to David Henderson and Stephen Devane for their artwork and to Peter Baker for his black and white drawings. The Meteorological Office at Bracknell have been very helpful in allowing me access to their archives.

Thanks also to the following people for taking the time to read the manuscript and their comments: Captain Brian Boardman (BP), Dr Frank Evans (Dove Marine Laboratory), Dr Peter Herring (Institute of Oceanographic Sciences), Dr Ronald Hope (The Marine Society) and Captain Hywell Phillips (BP).

A warm thanks to everyone who accepted me into their company at sea on board: mv *Vancouver Forest*, mv *Seatrain Saratoga*, mv *Troll Park*, mv *Wellpark*, mv *Manistee*, ss *British Trident*, ss *British Norness*, ss *Uganda* (on several occasions) mv *Author* and mv *Wellington Star*.

A special thanks to Bill and Eileen Westall for help, advice and encouragement in many ways.

Introduction

For as long as ships and boats have sailed the oceans, seafarers have watched the marine life that surrounds them. In the days of sailing ships, in calm weather there was little else to do except look over the side or to the horizon, see what was there, and pray for a fair wind. Today, ships are faster and rely little on the wind, but there is still time to gaze into the blue depths and even pull out a bucketful of water for a closer look. Over the years, seafarers have made very sharp observations, sometimes helped by a vivid imagination, and these have been included in the ships' meteorological logbooks. These logbooks contain accurate descriptions and drawings from which it is possible to identify the organisms.

For the last five years I have travelled on board British merchant ships collecting specimens and observing marine life. This book has been written mostly at sea as a result of my observations and the data accumulated in ships' logbooks. Many of the photographs and most of the drawings are of specimens collected or seen at sea, in ports and on beaches. It is hoped that this book will be interesting to people on the largest tankers, on passenger ships, fishing vessels, ocean-going yachts and small coastal craft. It is intended to enable anyone to identify any marine life they encounter in the course of their travels.

HOW TO USE THIS BOOK

All the species commonly seen are described. If a species is not in the text it will be possible at least to identify the group to which it belongs. The range of species likely to be seen is so wide that it is impossible to include all the possibilities. The appendix contains a list of books and organisations that will help the enthusiast in search of more information.

The first chapter is a general introduction to the marine world. Chapter two covers the range of small organisms you are likely to see through a magnifying lens or simple microscope. Chapter three looks at the strange phenomenon of bioluminescence, sometimes called phosphorescence. Each subsequent chapter is devoted to a particular group of species. At the start of each group there is a simple classification or 'family tree' covering the families in the group. This will guide the reader to the relevant section of the chapter. The organisms are then described individually, with brief notes on their particular habits.

Every species has a particular geographical range in which it lives, and at the back of the book there are distribution maps that will help in identification. In many cases, such as some of the mammals, it is possible to identify a species purely from where it is found.

The scientific names of each species are given together with any common names. This is to ensure that there is no confusion with other books and also because it is felt that they are of interest. With practice anyone can become fairly expert in the marine world; after all, the old whalers could identify different whales by the shape of their blow from miles away — why not today's seafarers?

Finally, the last chapter raises some questions about the future of our marine world. With 70,000 ships plying the oceans, and the subsequent increase in accidents; the increase in oil exploration, mineral extraction and fishery exploitation; the demands of tourism, and the need for education and conservation, what does the future hold? Can the marine world cope? Who has to take responsibility? These are some of the problems being faced today by governments, industry and seafarers.

1. The Ecology of the Sea

The marine world extends from the ocean surface to the deepest sea-bed and from shore to distant shore. There are many millions of animals and plants to be found in this world, but this book is a guide to those that are likely to be seen by seafarers, at or near the surface of the sea.

Ecology is the study of where living things are found, and why they live in particular areas. Animals and plants need certain conditions in order to live: oxygen and light are required in sufficient quantities; various kinds of nutrients are necessary; and the temperature must be favourable. Different species have different requirements. A species living in an area that has all the right conditions is said to be in its **habitat** (Figure 1.1).

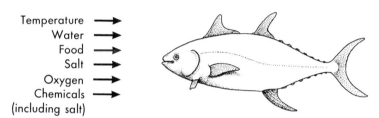

Temperature ➝
Water ➝
Food ➝
Salt ➝
Oxygen ➝
Chemicals ➝
(including salt)

Figure 1.1: Factors affecting marine organisms

VARIABLES IN THE MARINE ENVIRONMENT

The sea is not the same everywhere, but changes with different areas. These changes — in temperature, saltiness, amount of oxygen and so on — have an effect on the marine life. It is not possible to discuss here all the variations and their implications, but below are some of the most important. Although each is considered under a separate heading for clarity, in life they are inseparable, each affecting the others.

SALINITY (SALTINESS)

The average amount of salt in the sea is about 35 parts of salt per 1,000 parts of water. The body fluid of all marine invertebrates is at the same concentration as, or very near to, the concentration of the seawater. (The blood of land animals, including man, is about half the seawater concentration.) By contrast the body fluid of freshwater animals is more concentrated than the surrounding water; in these animals the water is constantly trying to enter and dilute the fluid to make it the same concentration as that outside. Freshwater animals must excrete water to maintain the higher concentration inside their bodies. Marine animals do not need to do this. Marine fish excrete salt to keep their fluid concentration lower than the seawater.

The salinity of the sea is not constant. Near land, fresh water dilutes the sea causing a lower salinity; near the poles and around icebergs melting ice also causes a lower salinity. But higher temperatures cause the water to evaporate, and this increases the salinity.

Coastal water species must be more tolerant to changes in salinity than oceanic water species, hence we can define organisms as being either coastal or oceanic.

TEMPERATURE

Temperature varies with latitude, and the oceans can be divided into a number of regions (Figure 1.2). The following divisions are often crossed as surface currents move between different areas. Warm currents may travel into temperate areas; the Gulf Stream, for instance, warms the west coast of Europe. Cold currents may travel to warm areas; for instance, the cold Humboldt current is felt as far north as the Galapagos Islands.

Equator to the tropics (23.5° north and south) — Equatorial or Tropical
The tropics to 40° north and south — Subtropical
40° north and south to 55° north and south — Temperate
55° north and south to 66° north and south — Circumpolar
66° north and south to the poles — Polar.

Organisms found over a large area must be able to tolerate large variations in temperature. Many whales migrate from tropical and subtropical waters, where they mate and breed, to polar waters where they feed. Whales are mammals and therefore are warm-blooded (meaning that they can maintain their internal body temperature regardless of the environment) and they can tolerate a wide range of temperatures. They have thick layers of blubber to keep them warm in the cold waters, but this is used up for energy on their journey to the tropics where food may be scarce.

All animals except birds and mammals are cold-blooded — that is, they have a body temperature that changes with the surrounding temperature. Some of these animals undertake long migrations too, but keep roughly to the same latitude. It is possible that some green turtles, *Chelonia mydas*, travel from one breeding site on Ascension Island in the tropical Atlantic to the Brazilian coast and back — a distance of 2,700 miles along the same latitude.

The combined effect of salinity and temperature changes the density of the sea, which in turn influences the ocean currents. These are discussed later.

LIGHT

Sunlight is vital for life; it is the major source of energy. This energy is used by plants to grow; the plants are then eaten by animals, and so the energy is passed on.

Plants do not need complicated food like carbohydrates and proteins: they can make their own from simple compounds. Carbon dioxide (CO_2), which is breathed out by all living things, and water (H_2O) are combined together, using the sunlight for energy, to make carbohydrates (which have the general formula $C_6H_{12}O_6$). This is the process of **photosynthesis**. Plants have special pigments, like the green chlorophyll, which are energised by the sun so the reaction can take place. The photosynthesis reaction can be written simply like this:

$$CO_2 + H_2O \xrightarrow{\text{sunlight and chlorophyll}} C_6H_{12}O_6 + O_2$$

carbon dioxide + water, with sunlight and chlorophyll gives carbohydrate + oxygen.

The carbohydrate is then used for food, and provides the energy for all the 'body functions': growing, building up new plant material and reproducing. Animals cannot make their own carbohydrates; they must either eat plants, or eat other animals that have eaten plants, in order to get their energy.

The above equation shows another vital job that plants and sunlight do. Photosynthesis produces oxygen, which enables all living things to survive: no plants means no oxygen.

In the sea the only obvious plants are the seaweeds that grow on shores or the *Sargassum* (gulfweed) found mainly in the north central Atlantic. Plants grow only in sunlight, and the depth to which the light penetrates in the sea varies with the clarity of the water. Around Britain seaweeds do not often grow below 18 metres (60 feet), whereas in clearer water, as in some parts of the Mediterranean, they may grow to depths of 30 metres (100 feet).

But what about the thousands of square miles of open ocean and its life? The large seaweeds are not sufficient to support the entire marine world. In the sea there are billions of tiny plants, visible only through a microscope (Chapter 2). These minute plants are part of the **plankton** — a floating collection of tiny organisms. Life in the sea would not be possible without the plant plankton.

The depth to which the sunlight penetrates marks the light zone; this is where the living plant plankton are found. In clear oceanic water the light zone extends to more than 150 metres (490 feet) and is the productive layer of the oceans providing all the energy for marine life. Deeper down is the dimly lit mid-zone extending to about 1,000 metres (3,280 feet), but the deepest water is in the abyssal zone — the black world of strange sea creatures (Figure 1.3).

A discussion of most of the abyssal animals is beyond the scope of this book, but there is much overlap because many animals migrate from one zone to another. Some midwater species come to the surface at night and give extensive displays of bioluminescence (Chapter 3); the reasons for this vertical migration are one of the mysteries of the sea. The lantern fish of the family Myctophidae are one example; they are found down to 500 metres (1,640 feet) during the day but at night they can be seen at the surface reflecting red and greeen 'cat's eyes' in a vessel's lights.

OTHER FACTORS

Oxygen is produced by the plant plankton in the light zone, hence the amount of oxygen depends on the productivity of the plant plankton which, in turn, relies on nutrients being available to the plants.

The organisms themselves change the physical and chemical properties of the water which affects other species. As an example, after winter the water temperature and amount of sunlight increases and there are nutrients from dead animals and plants that have accumulated during the colder months so the conditions are just right for the plant plankton. They increase and may grow to such numbers that often the colour of the sea changes (Chapter 2). This is a plankton 'bloom'. The number of animals also increases, because of the favourable conditions and abundant food. But as the nutrients and oxygen are used up, so the populations die and the cycle continues. This cycle of growth and death is an important feature of the natural world.

OCEAN CURRENTS

Figure 1.2: The major ocean currents and sea areas

The oceans are constantly moving, transporting animals and plants for thousands of miles. The surface currents are driven by the winds, and follow the prevailing wind directions which, due to the rotation of the earth, adopt roughly circular pathways (Figure 1.2).

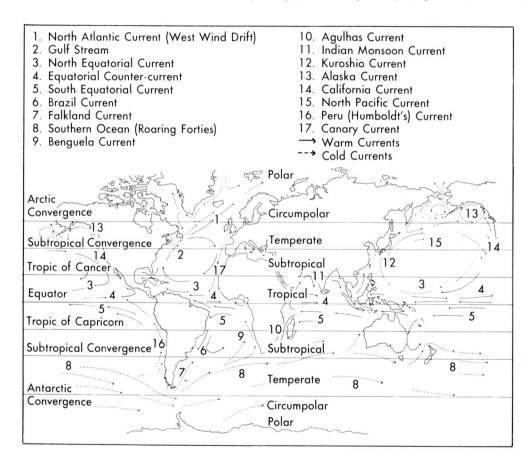

 1. North Atlantic Current (West Wind Drift)
 2. Gulf Stream
 3. North Equatorial Current
 4. Equatorial Counter-current
 5. South Equatorial Current
 6. Brazil Current
 7. Falkland Current
 8. Southern Ocean (Roaring Forties)
 9. Benguela Current
10. Agulhas Current
11. Indian Monsoon Current
12. Kuroshio Current
13. Alaska Current
14. California Current
15. North Pacific Current
16. Peru (Humboldt's) Current
17. Canary Current
⟶ Warm Currents
--→ Cold Currents

The density of seawater changes with its temperature and salinity. Cold water is heavier than warm, and salt water is heavier than fresh. Combining these properties, cold salt water is heavier than warm fresh water. In the oceans, when a cold current meets warm water it sinks — deep water is colder than surface water. The cold water may surface again perhaps thousands of miles from where it 'disappeared'

As different areas of water have different physical and chemical properties of salt content, temperature, and so on, and different animals require different conditions, it is possible to identify areas of water by identifying the organisms found in the sea. In this way scientists have been able to plot the movements of the currents.

Where water masses meet, and where currents are forced up near land or over submarine banks, nutrients are brought to the surface. These are known as upwellings. The plant plankton use the nutrients and, with the sunlight, multiply to create a rich source of food for other organisms. These areas are where the world's richest fishing grounds are found, such as those off Peru (although the Peruvian anchovy fishery has suffered from over-exploitation — see Chapter 12) and southwest Africa where the offshore winds blow surface water away from the land, and deep water is then pulled to the surface to form an upwelling. Other rich fisheries are found off Cape Hatteras on the east coast of the USA where the Labrador current meets the Gulf Stream; and among the Faeroe Islands where the west wind drift rises over a submarine ridge.

MARINE LIFE

All those marine organisms that swim poorly and drift with the currents are called plankton (from the Greek meaning 'wanderer'). Most plankton are tiny creatures seen clearly only through a microscope, but there are large planktonic organisms such as the Portuguese man-of-war, or the *Sargassum* weed. The plant plankton live at or near the surface but the animals are found at all depths. By contrast, animals such as fish, dolphins, whales and turtles are strong swimmers and often travel large distances against the currents. The sea-bed is the home for sliding, burrowing, walking or attached organisms, many of which be found on ships' hulls and in seawater pipes, coolers and condensers.

How do all these animals and plants — the seaweeds, plankton, fish and whales — work together to build the marine world? Plants are food for herbivorous animals, which in turn are eaten by hunting carnivores. This is a **food chain**. A simple food chain on land could be:

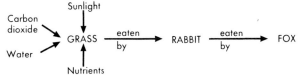

The grass uses sunlight, carbon dioxide and water, plus nutrients from the soil. The fox and rabbit put nutrients back into the soil in their waste material. When all living things die, their bodies decompose and their valuable nutrients are recycled by bacteria so that more grass can grow. This recycling is vital to the food chain, which is better shown like this:

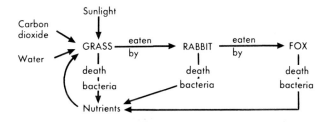

In real life rabbits are not just eaten by foxes; owls, buzzards, hawks and man also take their share. If these animals are included, the picture becomes a complicated web. This is the web of life — the **food web**.

In the sea, the web of life is slightly different:

The plant plankton can support more stages, hence the marine food chain may be longer than a land food chain (Figure 1.3). But in areas of great productivity, such as off the Peruvian coast, the food chains are shorter. When the animals and plants die they sink and provide food for the deeper animals, hence deeper waters are rich in nutrients, and it is these that come to the surface in areas of upwelling and mixing currents.

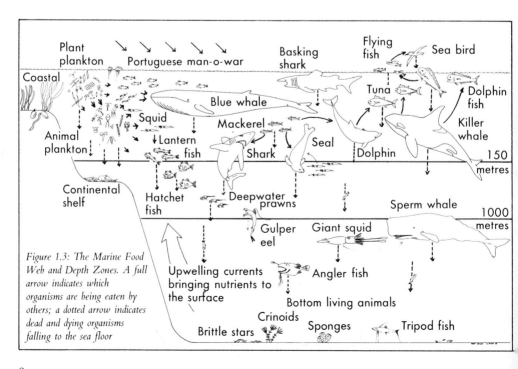

Figure 1.3: The Marine Food Web and Depth Zones. A full arrow indicates which organisms are being eaten by others; a dotted arrow indicates dead and dying organisms falling to the sea floor

The food web is intricate and delicate. If a disaster strikes — an oil spill, an industrial discharge, over-fishing — then the web suffers damage; nature must then readjust to make repairs that may take many years. We are part of the web and we must be careful of the damage that we can do: if the plant plankton is killed the marine world will break down and an important supply of oxygen will be lost; if there are few herring the cod take other food (sand eels and sprats); no cod and the seals take the herring; no killer whales and there is no control over the whole food chain.

CLASSIFICATION

The natural world is not a haphazard arrangement; the millions of animals and plants are related (see inside front cover). An internationally recognised system of classification is used to organise all the species and show the relationships between them.

Classification is rather like a family tree. The entire living world or **biosphere** (*bio* means life) is divided into two **kingdoms**: the animals and the plants. Each kingdom is divided into **phyla** (singular — **phylum**). The phylum is the major group heading used in this book, but there are other divisions: **class**, **order**, **family**, **genus** and **species**. Each phylum is made up of several classes; each class has several orders; and so on until the species defines one particular organism which has a two-part Latin name. In many cases there may be other groups added such as sub-phylum, sub-class, and sub-order but these have been mostly omitted in this book.

A **species** is a group of organisms that can interbreed to produce offspring that can themselves breed. A horse and a donkey are different species because although they are capable of interbreeding, the offspring they produce — the mule — is sterile.

Each division of the classification has a set of characteristics that apply only to the members of that group. Generally, animals do not have chlorophyll; plants do; this is the first division. At each stage the characteristics are matched to the species that is being considered (Figure 1.4).

In figure 1.4 the family **Velellidae** has only two genera (plural of genus). The classes are related: in this case they are all jellyfish-like **coelenterates**. The orders are more closely related — they are anemone-like **hydrozoans**. And the relationships become closer until the by-the-wind-sailor, *Velella velella*, is very closely related to *Porpita porpita* (Chapter 4 discusses these animals in detail).

This system of classification is used throughout the world, so although the common names of species may vary from place to place the Latin name is always the same.

After the generic and specific names of an organism it is common practice to add the name of the person who first described it, e.g. *Ostrea edulis* Linnaeus is the common edible oyster and was first described by Linnaeus. Many species have Linnaeus' name, reflecting the amount of work done by this scientist. If the name is in brackets e.g. *Velella velella* (Linnaeus), it indicates that although Linnaeus was the first to describe the animal, further research has put it in a different group. In the following chapters the names of the scientists have been omitted for the sake of simplicity.

Figure 1.4: The classification or 'family tree' of Velella and Porpita

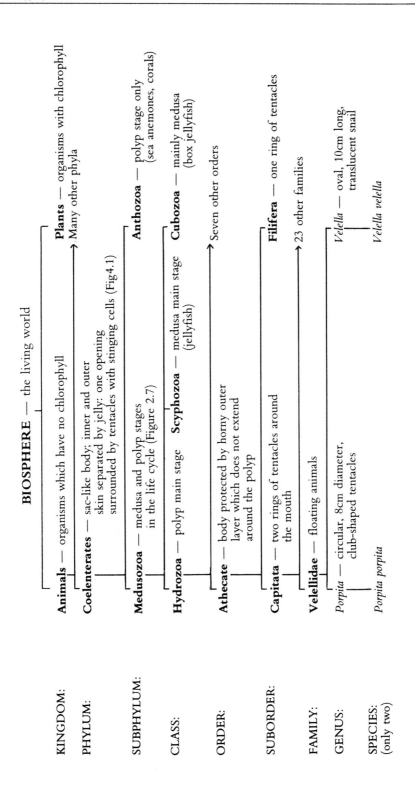

BIOSPHERE — the living world

KINGDOM:
- **Animals** — organisms which have no chlorophyll
- **Plants** — organisms with chlorophyll
  → Many other phyla

PHYLUM:
- **Coelenterates** — sac-like body; inner and outer skin separated by jelly; one opening surrounded by tentacles with stinging cells (Fig4.1)

SUBPHYLUM:
- **Medusozoa** — medusa and polyp stages in the life cycle (Figure 2.7)
- **Anthozoa** — polyp stage only (sea anemones, corals)

CLASS:
- **Hydrozoa** — polyp main stage
- **Scyphozoa** — medusa main stage (jellyfish)
- **Cubozoa** — mainly medusa (box jellyfish)

ORDER:
- **Athecate** — body protected by horny outer layer which does not extend around the polyp
  → Seven other orders

SUBORDER:
- **Capitata** — two rings of tentacles around the mouth
- **Filifera** — one ring of tentacles

FAMILY:
- **Velellidae** — floating animals
  → 23 other families

GENUS:
- *Porpita* — circular, 8cm diameter, club-shaped tentacles
- *Velella* — oval, 10cm long, translucent snail

SPECIES:
(only two)
- *Porpita porpita*
- *Velella velella*

2. The Miniature World

Two-thirds of the world's surface is under water. In some areas a bucketful of seawater will contain several hundred thousand tiny plants and thousands of animals. This is the life-source of the oceans — a world visible only through a magnifying lens (Black and White Plate 1).

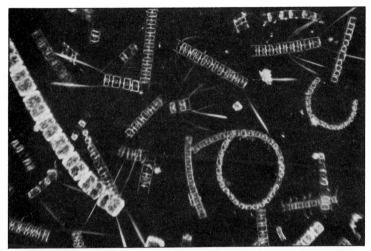

Plate 1a. A collection of microscopic plants or **phytoplankton**

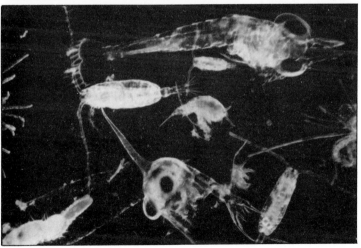

Plate 1b. A collection of microscopic animals or **zooplankton**

These organisms are part of the plankton. The plants, or **phytoplankton**, range in size from a few microns to several hundred microns (1 micron is one thousandth of a millimetre — a human hair is about 50 microns in diameter). The animals — **zooplankton** — are larger, ranging from about 50 microns to a few centimetres.

Permanent plankton are those organisms that spend their whole lives drifting, but the eggs and young of many larger animals are cast adrift in the sea, and these are temporary plankton. Crabs, lobsters, limpets, mussels, herring and plaice are just some of the animals that start life as plankton. The sea is an efficient way of spreading the species to other areas.

Seaweeds are plants without flowers: they belong to the group called **algae**. Male and female spores are shed into the water where they combine (fertilisation) and new seaweeds start to sprout. If the young seaweeds are to survive they must be washed to suitable places where they can attach themselves and grow.

CHANGES IN THE AMOUNT OF PLANKTON

Coastal water contains nutrients swept up from the sea-bed by upwelling currents and poured into the sea from rivers. These nutrients, for example nitrate and phosphate, are used by the phytoplankton to grow, and so provide food for the zooplankton which contains many offspring from animals of the shore and sea-bed. The concentration of plankton and nutrients gives coastal water that grey/green colour which is so different from the deep blue of the oceans.

As you sail away from the coast the amount of temporary plankton decreases until, on the high seas, the occasional fish eggs or young fish and the young of other oceanic animals are all that is found. Oceanic water has relatively more permanent plankton, but generally the total volume of plankton in the open sea is less than on the coast.

Not only does the amount of plankton depend on the area (Map 1) but also on the season and the time of day. On land we recognise four seasons. The sea, too, has seasons although they follow a slightly different pattern. During the winter many plants and animals die and their bodies, which rot and break up, add to the nutrients in the water. As days get longer and the temperature rises the phytoplankton grow quickly and provide food for the zooplankton which consequently also increase in number. This is spring, the growing season, when the populations are at a peak. In summer the food is used up and the numbers fall slightly but in autumn there is another burst of growth as autumnal gales stir up the water bringing nutrients to the surface. As winter approaches days get shorter (hence the amount of light falls), the numbers drop to the winter low and the cycle prepares to repeat (Figure 2.1).

Near the poles there is a short increase in growth during the brief summer season. But as you approach the equator the plankton becomes scarce because there is no winter 'resting' period; the growing season lasts all year so the nutrients are continually being used. Consequently tropical fisheries are poor except in waters influenced by upwellings or river estuaries.

The richest areas for plankton are upwelling areas and where strong winds stir up the waters, for example around the Antarctic. Not only are there abundant nutrients due to the mixing of currents, but the short summers of constant sunlight also cause an enormous bloom of phytoplankton to feed the large amount of animal life. These are the areas to which the great whales migrate to feed. 'Krill' is the name given by the whalers to the large zooplankton found in the polar areas (Chapter 2, p. 36).

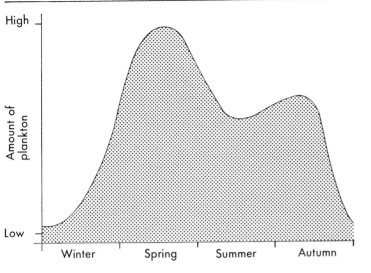

Figure 2.1: Variation in the amount of plankton throughout the year

THE VERTICAL MIGRATION MYSTERY

Most zooplankton can swim, but they are too small to move against currents. Yet their swimming ability enables them to travel up and down in the water, migrating from deep water to the surface and back each day. This is the phenomenon of **vertical migration** (Figure 2.2). The distance travelled varies with different species. Some of the larger crustaceans such as shrimps and krill may travel between depths of 600 to 1,000 metres (1,970 to 3,280 feet) and the surface. The smaller crustaceans travel between 30 and 150 metres (100 and 500 feet) and the surface.

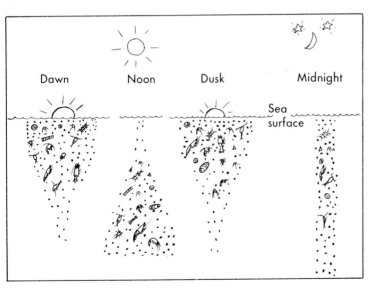

Figure 2.2: Vertical migration in plankton

At dawn many animals are at the surface but, as the sun rises, they sink and are back at their deepest level by midday. Towards sunset they swim back to the surface, but overnight they arrange themselves randomly.

The reasons behind this behaviour are a mystery. Many factors are important. It is known that some species migrate while close relatives stay in the depths. By swimming to the surface and leaving the other animals behind there is less competition for food either at the surface or in the depths. Vertical migration also enables the animals to have some control over where the oceans are carrying them. Deeper water currents often move in quite different directions to the surface currents. If there is an east-west current at the surface and a west-east current in deeper water, at dawn the surface animals will be carried westward. By midday the animals will have sunk to depths where the water carries the animals eastward; then by swimming back to the surface they will, once again, be carried westward. In this way the animals execute a roughly square course and so maintain the same position, relative to the sea-bed, for days or weeks. Because of vertical migration, generally the best time to catch plankton is either at dawn or dusk.

HOW TO COLLECT AND STUDY SMALL MARINE LIFE

EQUIPMENT

Specimens of plankton have been obtained by dropping a bucket on the end of a rope over the side of the ship, as when taking seawater temperatures, but this is only useful in areas where the plankton is very rich.

Ready-made nets are expensive, but it is possible to buy net material of any particular mesh size (about 200 microns is a good size) and make your own. The net should be conical, with the narrow end diameter slightly more than the diameter of the sample jar, which can be fastened with a string or jubilee clip. At sea, plastic jars with screw-on lids are preferable, but glass jars are used for long-term storage. Small plastic dishes are used to hold the specimens while you look at them. A magnifying glass or small hand-lens magnifying by 10 or 15 times, is sufficient to examine many of the organisms in the sample. To see more detail a microscope must be used — a simple, inexpensive, monocular student microscope is sufficient. The small Open University MacArthur microscope is compact, robust, easy to use and ideal for use at sea.

Plate 2 collecting plankton through a ship's fire hydrant

COLLECTING SAMPLES

The best way to collect these tiny creatures is to tow a fine mesh net slowly (2 to 4 knots maximum) through the water. This is usually impossible from a merchant vessel under way, but the same effect can be achieved at anchor by letting the net drift out with the current. At sea, specimens can be obtained by running seawater from the deck hydrant through the net (Black and White Plate 2).

The time needed to get a good sample will depend on the method used, where you are, and the time of day. At anchor you will be near the coast, and towing the net for 10 or 15 minutes will be sufficient — probably less at dawn or dusk. But at sea, using the seawater pump, more time is required; one or two hours at dawn or dusk usually gives an interesting sample although some of the larger animals will have been broken passing through the pump.

After the water has passed through the net for a specific time, remove the jar but have a bucket of water handy to wash off any stray animals left clinging to the net. Look at the sample for any unusual or obvious plants or animals, then put a label in the jar — a label on the outside can become detached. Note the date, time and position of the ship before and after the sample was taken using a soft lead (HB) pencil on good quality paper to prevent the label disintegrating and becoming illegible. This label will keep for years.

If you wish to view the specimens alive the sample should be looked at immediately — the plankton will not survive long in a small jar. If the sample is stored in a fridge, many animals can be kept alive for a few days; they become inactive as the temperature falls and may appear dead on the bottom of the jar, but as the water warms up their activity increases. It is not advisable to keep the samples too long before preserving them, and without a fridge the sample should be preserved within a few hours.

STUDYING THE SPECIMENS

Put a few specimens in a dish with some water using a teat pipette or eye-dropper, then look at them through the hand-lens or magnifying glass. To use the lens hold it about an inch from the eye and bring the subject into focus, don't move the lens up and down to focus on the subject. The magnifying glass is easier to use but does not show up as much detail.

Once you have selected a particular specimen to investigate, remove it in a drop of water with the pipette or dropper and place it on a glass slide (Figure 2.3). Put four small pieces of plasticine on the corners of the cover slip, drop the slip over the specimen and gently press on the four corners. The slide is ready to view. An alternative is to use a ring of plasticine which prevents the specimen from drying up over a period of about a week.

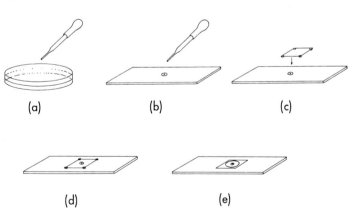

(a) (b) (c)

(d) (e)

It should be possible to identify roughly the types of specimens in your sample from the notes in this chapter. Once you have decided what the sample contains, make a list of them together with a clear indication of which sample they belong to, or write them on another piece of paper to be put in the jar.

Figure 2.3: Preparing a slide mount

The organism is removed in small drop of water with the pipette (a); then it is transferred to the microscope slide (b). Small pieces of plasticine are attached to the corners of a cover slip which is gently lowered onto the slide (c); gentle pressure on the corners of the coverslip will manipulate the organism to achieve the best view (d). If the slide is to be kept for a few days a ring of plasticine must be used to stop the specimen from drying out (e).

PRESERVATION

Alcohol is the best material for preserving specimens, but this is not generally available. Ordinary methylated spirit or denatured alcohol is not suitable. Liquors such as gin may be used — but only in an emergency. The most practical preservative is formalin, which can be bought at a chemist. Make up a ten per cent solution by adding nine volumes of seawater to one volume of formalin. Formalin must be handled with care as it causes intense irritation to eyes, nose and cuts, and if spilled in a confined area it can be serious. It is best to use it in a well-ventilated room. Add a few lumps of chalk or coral pieces otherwise the formalin, which is slightly acidic, will dissolve any skeletons and shells leaving some animals unrecognisable. Pour off as much of the seawater from the sample as possible without losing the animals and refill with the same amount of diluted formalin.

Several organisations will be interested in your specimens and will give you more information about them. The Meteorological Office at Bracknell deals with most of the enquiries from ships. You should send the specimens with the ship's meteorological logbook when it is returned to Bracknell. If you prefer to pursue the investigations yourself the Natural History part of the British Museum is a useful source of information.

IDENTIFICATION

In the miniature world there is no absolutely clear distinction between animals and plants; some organisms have animal character-istics *and* the green plant pigment chlorophyll. These organisms can make their own food in the same way as ordinary plants, but they can also digest other plants and animals. For our purposes those organisms with chlorophyll are plants and those without are animals.

It is impossible here to name all the specimens that are likely to be caught, but using the following notes and Figure 2.4 it should be possible to identify the major groups. Figure 2.4 is a quick key to the major types of organisms that may be found in this miniature world, but for complete identification, specialist books are required in most cases.

THE PLANT KINGDOM — PHYTOPLANKTON

A net with a very small mesh size of between 100 and 150 microns is required to catch the phytoplankton. These plants are minute but fascinating, showing, through the microscope, a world of intricate geometrical patterns (Figure 2.5). Many live in cases made from silica or calcium carbonate which, over millions of years, help to form chalk deposits on the sea-bed. The white cliffs of Dover are made from billions of animal and plant plankton cases.

Living plants must keep themselves in the lighted zone of the sea; some have long spines and oil droplets (oil floats on water) which prevent them from sinking — these are **diatoms**. (Many animals and plants possess oil which is released when their bodies are smashed by breaking waves and forms a froth — **spume** — on the water.) Other plants have long whip-like tails that beat to move them through the water. The tail is called a **flagellum** and the group are the **flagellates**.

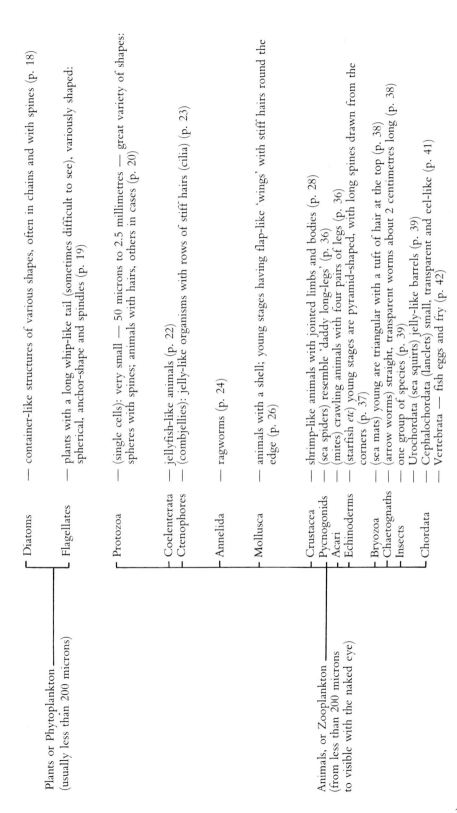

Figure 2.4: A simplified guide to the miniature world

Plants or Phytoplankton (usually less than 200 microns)

Diatoms — container-like structures of various shapes, often in chains and with spines (p. 18)

Flagellates — plants with a long whip-like tail (sometimes difficult to see), variously shaped: spherical, anchor-shape and spindles (p. 19)

Protozoa — (single cells): very small — 50 microns to 2.5 millimetres — great variety of shapes: spheres with spines; animals with hairs, others in cases (p. 20)

Coelenterata — jellyfish-like animals (p. 22)

Ctenophores — (combjellies): jelly-like organisms with rows of stiff hairs (cilia) (p. 23)

Annelida — ragworms (p. 24)

Mollusca — animals with a shell; young stages having flap-like 'wings' with stiff hairs round the edge (p. 26)

Animals, or Zooplankton (from less than 200 microns to visible with the naked eye)

Crustacea — shrimp-like animals with jointed limbs and bodies (p. 28)

Pycnogonids — (sea spiders) resemble 'daddy long-legs' (p. 36)

Acari — (mites) crawling animals with four pairs of legs (p. 36)

Echinoderms — (starfish *etc*) young stages are pyramid-shaped, with long spines drawn from the corners (p. 37)

Bryozoa — (sea mats) young are triangular with a tuft of hair at the top (p. 38)

Chaetognaths — (arrow worms) straight, transparent worms about 2 centimetres long (p. 38)

Insects — one group of species (p. 39)

Urochordata (sea squirts) jelly-like barrels (p. 39)

Cephalochordata (lancelets) small, transparent and eel-like (p. 41)

Vertebrata — fish eggs and fry (p. 42)

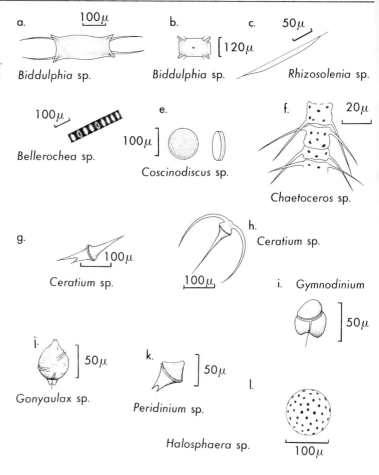

Figure 2.5: A selection of miniature plants — phytoplankton

a to f are diatoms (a and b are two species of Biddulphia*); g to k are dinoflagellates (g and h are two species of* Ceratium*); l is a yellow-green alga.*

Diatoms

The sizes range from a few microns to about 1 millimetre, but chains of individuals may be several centimetres long. Each plant lives in a silica case, which is in two parts forming a lid and base like a pillbox. Some have patterns radiating from one or more points — these are **centric diatoms**; others have patterns in straight lines similar to a feather and are called **pennate diatoms** (pennate means feather-like). Pennate diatoms generally live on the sea-bed near the coast, hence centric diatoms are the most common in plankton samples.

Three typical centric diatoms are:

Coscinodiscus — simple disc-shaped cells between 50 and 100 microns in diameter.

Rhizosolenia — cells about 300 microns long with a point at each end which may be sharp or blunt. Plankton-feeding fish like herring avoid patches of some species of *Rhizosolenia* which impart a yellow colour and bitter taste to the water. Fishermen call such areas 'baccy juice water'.

Chaetoceros — oval cells, about 15 to 70 microns in diameter, with flat ends and a pair of long thin spines. Many cells may interlock making chains several hundred microns long.

Navicula is the most common pennate diatom; it is found in most coastal waters. The cell is about 100 microns long and shaped like a boat with a rounded bow and stern. Usually these diatoms are picked up in coastal plankton samples after stormy weather when the sea-bed will have been stirred up.

Flagellates

These are borderline between animals and plants, and many books classify them as animals even though some have chlorophyll. All these organisms have at least one tail or flagellum.

The dinoflagellates, which are the main group, have two flagella; one trailing in a groove running lengthways, the other in a groove across the organism.

Four common dinoflagellates are:

Ceratium — perhaps the most characteristic. The cell is flattened and the case has two or three horns 50 to 300 microns long. The flagella are wrapped around the centre.

Peridinium — the commonest dinoflagellate, resembles *Ceratium* without the long horns. It looks like a spinning top about 50 microns in diameter.

Gymnodinium — has an egg-shaped case about 50 microns in diameter and is commonly found in coastal water.

Gonyaulax — is also about 50 microns in diameter and looks like a rounded spinning top.

Red tides

Gymnodinium and *Gonyaulax* are two species that are responsible for the phenomenon known as a 'red tide' which occurs when large numbers of animals or plants become concentrated in one area. This happens when the organisms are caught in a region where two currents meet to form a downwelling — that is where the water is sinking. A combination of the organisms swimming towards the surface and the water moving downwards results in a concentrated population. Often these are seen in long lines, streaks or windrows on the surface. The water currents carry nutrients and the organisms to potential areas so a red tide may persist for a week or two. The Red Sea is so named because of the frequent appearance of red tides, but they occur all over the world (Map 2).

In a red tide the water often has the appearance of rust or paint, but the colour may vary according to the species responsible with shades of red, pink, violet, orange, yellow, blue, green or brown being the most common recorded.

When certain species of *Gymnodinium* and *Gonyaulax* are in such concentrations that they form red tides they are responsible for the mass killings of many animals. This happens in two ways. Firstly, when the plants have used up all the nutrients, or the temperature changes, or the amount of sunlight falls, then the entire population dies simultaneously. As the plankton decompose, the amount of oxygen in the water falls so that many other animals 'suffocate'. In 1977 a red tide off New Jersey, USA, caused an area of over 14,000 square kilometres (5,400 square miles) to become deficient in oxygen with the result that many animals were killed, especially those living on the sea-bed like clams and mussels which could not escape.

The second way these plants can kill is by producing poisonous substances, some of which kill fish while others are more effective against warm-blooded animals. The west coast of Florida is known for frequent and extensive fish kills caused by red tides. The organism responsible is *Gymnodinium breve*, which gives the water an oily appearance and produces a poison that causes red blood cells to burst. The organism is fragile and breaks as it passes through the gills releasing the poison which is absorbed by the blood; the burst blood cells are unable to carry oxygen and the fish dies gasping on the surface. Florida's tourist trade has suffered because piles of dead fish have been washed up along the shores and people exposed to wind-blown spray from red tide areas have suffered from respiratory irritations.

Another poison, produced by the *Gonyaulax* group, is responsible for the disease known as 'paralytic shellfish poisoning'. Warm-blooded animals are particularly sensitive to this poison which affects the nerves and leads to muscular paralysis and death. Some cold-blooded animals like fish are affected but shellfish, like clams and mussels, are unharmed and the poison accumulates in their bodies. In extreme cases if just a few infected shellfish are eaten by man they may prove fatal. Unfortunately poisoned shellfish appear normal so there is no indication of potential danger.

Records of paralytic shellfish poisoning are world-wide. Most outbreaks have occurred along the west and northeast coasts of North America, the coasts of northwestern Europe and the British Isles, around Japan, along the coast of British Columbia, in the southern Bay of Fundy and in the St. Lawrence river estuary. The annual bloom of *Gonyaulax* species usually means that certain areas are closed to shellfish harvesting during that time of the year.

A second group of flagellates comprises the **coccoliths**, which are about 50 microns in diameter and made from tiny calcium carbonate plates. Billions of these cells are primarily responsible for chalk deposits. Two of the most common are:

Halosphaera viridis — a round yellow-green alga about 150 microns in diameter.

Phaeocystis — a brown flagellate that makes a slimy covering when it occurs in large numbers.

THE ANIMAL KINGDOM — ZOOPLANKTON

Much of the miniature zooplankton consists of the offspring of larger animals, but there are adults that do not grow to more than a few millimetres. The animals are discussed in order, from the simplest, made of a single cell, through to those with many cells. Most of the zooplankton are larger than the phytoplankton so a net with a mesh size of about 250 microns is sufficient to catch a good sample.

The protozoa — single-celled animals

There are many varieties of tiny single-celled animals at all depths of the ocean, and even among the sand grains of the sea-bed. They range from 50 microns to about two and a half millimetres (Figure 2.6). Three groups can be found in the sea:

the **flagellates** have mostly been discussed; the **rhizopods** are like blobs of jelly inside a prickly ball-like shell; and the **ciliates** have short stiff hairs.

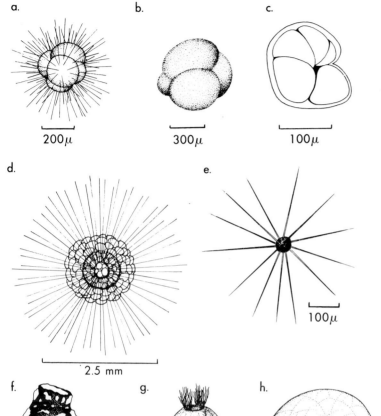

a.

200μ

b.

300μ

c.

100μ

d.

2.5 mm

e.

100μ

f.

50μ

g.

100μ

h.

500μ

Figure 2.6: A selection of single-celled animals — zooplankton

Foraminifera: a and b are two species of Globigerina. *Radiolaria: d is* Thalassicolla *spp. Acantharia: e is* Acanthochiasma *spp. Ciliates: f is a tintinnid vase-like case and g is an ologotrich. Dinoflagellate: h is the bioluminescent* Noctiluca.

One species of flagellate that does not have chlorophyll is a giant dinoflagellate called *Noctiluca scintillans* which grows to a millimetre or more in diameter. It is a jelly-like ball with a flagellum and a small tentacle extending from one side. When they are concentrated in large numbers they are responsible for dramatic displays of bioluminescence (Chapter 3).

Flagellates

The word 'rhizopod' actually means 'extending foot', and most of these animals move by pushing out a foot-like extension and dragging the body behind. The simple *Amoeba* found in ponds and lakes, is a typical example. Most of the species found in marine plankton samples live inside a case with holes through which parts of the body extend, usually along long spines so

Rhizopods

they look like small prickly balls. Some species are brightly coloured, either red or yellow. The most common groups are:

Foraminifera — have shells made of calcium carbonate (limestone). In warm areas millions of these shells form muddy sea-bed deposits called **calcareous ooze**. One common group of species is *Globigerina* which is about 400 microns in diameter. Foraminifera shells are also responsible for chalk and limestone cliffs.

Radiolaria — tiny animals with shells made from silica, a clear material also used commercially for making glass. They live in colder seas where they form deposits of **siliceous ooze** on the sea-bed. *Thalassicolla* is one of the largest with a diameter of two and a half millimetres.

Acantharia — have spines made from strontium sulphate which meet at the centre forming a regular pattern. *Acanthochiasma* has a body of about 150 microns in diameter with spines about 200 microns long.

Ciliates
These animals swim by using short stiff hairs called **cilia** which beat in rhythm like hundreds of tiny flapping wings. Around the mouth are longer hairs which are used to trap particles of food.

Oligotrichs have only a few cilia, which look more like bristles, and they do not have a case.

Tintinnids live in a case that is shaped like a vase and is made from particles such as diatom cases, glued into a cement-like protein substance. The head, with the cilia, projects out of the top of the 'vase'. Usually only the shells are caught in plankton samples.

The coelenterates — jellyfish, sea firs, anemones and corals

A diagram of the generalised life-cycle of a coelenterate is shown in Figure 2.7 (see also Figure 4.5). The jellyfish (**medusa**) stages are either male, producing sperm, or female, producing the eggs which hatch into a young stage called the **planula**. The medusa is planktonic but the polyp is attached to the sea-bed or to floating material.

The animals are classified according to which stage in their life-cycle lasts longer (Chapter 4). Sea firs, anemones and corals live mainly in the polyp stage; jellyfish live mainly as medusae. In some, evolution has made either one of these stages redundant; for instance anemones or corals have no medusa stage but the polyps produce the eggs and sperm which develop into a planktonic young. Similarly, some jellyfish have no polyp stage.

Miniature jellyfish caught in plankton samples are the reproductive (medusa) stages of sea firs. They produce the eggs and sperm which, in some species, are shed into the water, but in others the eggs are fertilised inside the body of the medusa. Adult sea firs attach themselves to the sea-bed, drifting debris, pier piles, ships' hulls and seawater pipes.

All coelenterates have a planktonic young stage, the planula, which looks like a tiny hairy sausage and lives from a few hours to a couple of days before changing into an adult. They may be caught at certain times of the year near the coasts where the adults are found.

The medusa is umbrella-shaped and generally between 2 and 100 millimetres in diameter, though some grow to

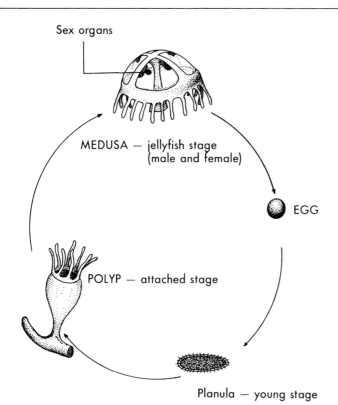

Sex organs

MEDUSA — jellyfish stage
(male and female)

EGG

POLYP — attached stage

Planula — young stage

15 centimetres. Tentacles bearing numerous stinging cells (Chapter 4 and Figure 4.1) grow around the edge, and the mouth is in the centre under the umbrella (Colour Plates 2, 3 and 4). Light-sensitive organs are located at the base of each tentacle. Some species swim towards the light and are found near the surface during the day; others swim away to deeper water during the daylight. There are also organs sensitive to gravity. These consist of little chambers, each with a grain of limestone inside that sits on sensitive nerve hairs. It is like a ball in a bowl; when the animal is the right way up the grain is on the bottom, but when the animal is upside down it rolls to the top. The balance organs in our inner ears work on the same principle.

The ctenophores — sea gooseberries, comb jellies or sea walnuts

Most sea gooseberries are like small round lumps of jelly ranging from about the size of a pea to that of a golfball, though there are other shapes (Figure 2.8). They have clumps of stiff hairs set in eight rows, each clump resembling a comb (the Greek *ktenos* means a comb — hence the group's name). The hairs beat in rhythm, so enabling the animal to move through the water.

Sea gooseberries do not have stinging cells like the jellyfish; small animals are caught by sticky cells on the two long tentacles and then swung round into the mouth.

These animals are found world-wide. One species common in temperate water is *Pleurobrachia* (Black and White Plate 3). Another species, *Cestum*, is flattened, grows to over a metre long and is commonly called Venus' Girdle.

23

Figure 2.8: Three species of comb jelly, or ctenophore

Cestum veneris — Venus' Girdle
(up to 1.5 metres)

Bolinopsis spp.
(up to 15 cm)

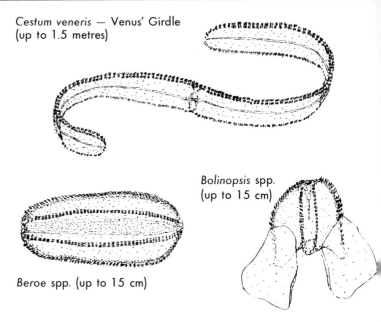

Beroe spp. (up to 15 cm)

Plate 3: The ctenophore, or sea gooseberry, Pleurobrachia *sp.*

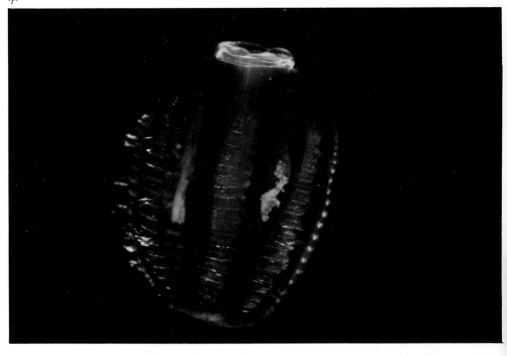

**The annelids —
ragworms and
tubeworms**

These worms grow in segments resulting in rings around their bodies (the Latin *annulus* means a ring). Ragworms are usually found burrowing in the sand and are often dug up for fish bait, but tubeworms live in little white limestone tubes stuck to rocks, seaweed, ships' hulls and seawater filters. Both types of worm produce planktonic young that can usually be found in coastal water.

The young stage of a worm is egg-shaped and called a

trochophore. It has a tuft of hairs at the top and bottom and one or two rows of hairs around the middle (Figure 2.9). As the young grows, segments are added to the bottom until it begins to resemble the adult, then it sinks to the sea-bed to grow into a mature worm (Black and White Plate 4). Some tubeworms keep the young in their tubes (brooding) until they have grown a few segments; then they are released into the water.

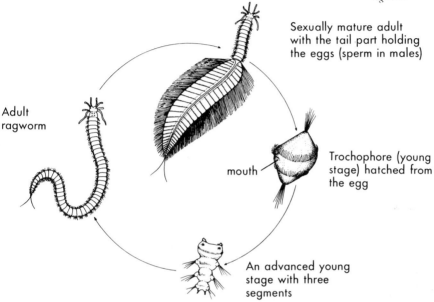

Figure 2.9: The life-cycle of a ragworm

Sexually mature adult with the tail part holding the eggs (sperm in males)

Adult ragworm

mouth

Trochophore (young stage) hatched from the egg

An advanced young stage with three segments

Some ragworms reproduce dramatically: at sexual maturity their internal organs disintegrate and the body becomes swollen into a bag, the female full of eggs; the male full of sperm. When the time is right the worms swarm to the surface, their bodies burst and the eggs and sperm are shed into the sea. Often only the tail part of the worm is involved (Figure 2.9): this swells, breaks free and swims to the surface. The remaining segments regrow the tail.

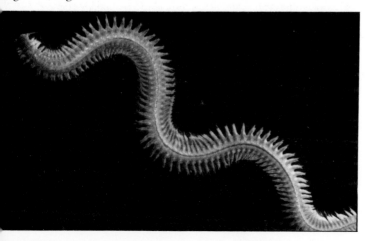

Plate 4: A ragworm swimming

The palolo worm, *Eunice*, offers the best example of this kind of sexual behaviour. The Atlantic palolo is a West Indian worm that normally lives in rock and coral crevices, but breeding swarms occur in July during the first and third quarters of the moon. At three or four o'clock in the morning the sexual part breaks free from the worm and swims to the surface where it performs spiralling movements. By dawn the surface is covered with these sexual bodies which burst as the sun rises, spreading eggs and sperm over a wide area. A day after fertilisation the young develops, and three days later the juvenile worm sinks to the sea-bed to grow into an adult.

The Samoan palolo worm swarms in October and November at the start of the last quarter of the moon, and is a local delicacy; during the swarm, thousands may be collected by the locals. The worms also provide a feast for any waiting fish or birds.

Although this behaviour is related to the moon's quarter, it is actually a result of the movement of the tide which in turn is related to the phase of the moon (Chapter 12). The presence of light — even the full moon — will prevent the swarming.

Tomopteris is a ragworm, 15 to 20 millimetres long, that spends its entire life drifting in the plankton. It has swimming legs like paddles down both sides of the body, and long feelers sticking out from the head. The worm is carnivorous, feeding on other planktonic animals including some small fish. When *Tomopteris* is under threat it rolls into a ball, like a hedgehog, and sinks out of harm's way. *Tomopteris* is transparent in life but becomes opaque when preserved.

The molluscs — snails, mussels, squid and octopus

Periwinkles, limpets and whelks (gastropods); mussels, scallops and oysters (bivalves); and squid, cuttlefish and octopus (cephalopods), are all molluscs. In all three groups there are species that produce planktonic young but, except for some rare bivalves, the gastropods are the only group with planktonic adults (Chapter 4).

Life starts with the egg. This hatches into a young stage or **trochophore** (which is similar to that of the ragworms, showing a distant relationship between these two groups). The trochophore quickly develops into an advanced stage called a **veliger**. However, most molluscs pass through the young stage inside the egg, which therefore hatches directly into the veliger (Figure 2.10). The octopus group (cephalopods) have no young stages; they develop directly into miniature adults. Most young molluscs are caught in coastal waters.

It is impossible to identify what type of mollusc a trochophore will develop into, in fact it is difficult to tell whether an early trochophore is a mollusc or young worm. But the veliger is very characteristic; at this stage it is possible to identify the type of mollusc it will grow into.

Usually the veliger has two, sometimes more, flap-like wings with hairs around the outer edge (Figure 2.10). When swimming, the hairs beat in rhythm and appear like a system of rotating wheels with spikes around the edges. When the veliger is first caught, the flaps are pulled into the shell so the animal looks just like a tiny snail or mussel but, if it is not disturbed, it will soon relax and start swimming. After the mollusc has developed all the adult structures the flaps are cast off so that the animal sinks and settles into its adult life.

Figure 2.10: The life-cycles of some molluscs

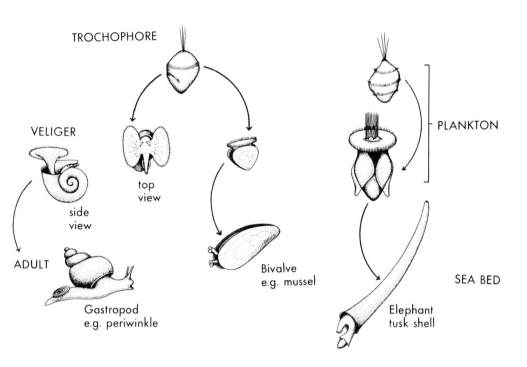

TROCHOPHORE

VELIGER

top view

side view

ADULT

Gastropod e.g. periwinkle

Bivalve e.g. mussel

PLANKTON

SEA BED

Elephant tusk shell

All gastropods begin life with the typical coiled shell. The snails keep the shell throughout their lives, but the slugs cast it off at some time during their growth and development. The flaps keep the young gastropod afloat for an average of two to three weeks before they settle as adults. Those snails that spend their whole lives drifting (chapter 5) have a veliger stage that is usually found well out to sea.

Gastropods — snails and slugs

These animals have a shell in two halves, hinged together. Young bivalves have a light, clear, delicate shell but, as the animal grows, it becomes thicker and heavier so that more flaps develop to carry the extra weight. Eventually the animal sinks to the sea-bed and grows into an adult.

Bivalves — mussels, oysters etc.

These are found world-wide in soft sea-beds. The veliger has a round flap rather like a dinner plate sitting above the animal.

Scaphopods — elephant tusk shells

Young octopus and squid are easily recognisable; they look like miniature adults. After hatching they are about three millimetres long (one-eighth of an inch) but grow to about one and a half

Cephalopods — octopus, squid, etc.

centimetres (five-eighths of an inch) in the plankton before settling on the sea-bed. Coastal waters may, at certain times of the year, contain many young octopus.

The crustaceans — barnacles, crabs, shrimps and lobsters

The Latin *crustaceus* means 'hard outer skin'. All the animals in this group have an external skeleton, or **exoskeleton**, which must be cast off periodically and replaced with a larger one to enable the animal to grow. When the time is right, the animal hides in a crack or burrow, the old skeleton splits, and the animal emerges clothed in a new soft skeleton. This skeleton swells to the new size and hardens — a process that may take from a few hours to several days. During this period the animal stays in hiding because it is vulnerable to attack from predators. This moulting occurs often in the young but becomes less frequent as the animal grows older. Bits of moulted crustacean skeletons are often found in plankton samples.

Crustaceans develop from an egg which, in many cases, hatches into a young stage, the **nauplius**. This is small and kite-shaped, with an eye and three pairs of hairy legs. It grows through a series of moults (usually six), gradually getting more complicated until an advanced stage is reached. The advanced young continues through further moults until it develops into an adult. Some crustaceans pass through all these steps whereas others brood the eggs and release the advanced stage or even young adults.

About 90 per cent of the animal plankton are crustaceans. In fact only three groups of crustaceans have non-planktonic adults; the barnacles (cirripedia), the mantis shrimps (stomatopoda), and the crabs and lobsters (decapoda). But all these animals produce planktonic young. Most of the remaining groups have members that are entirely planktonic.

Cirripedia

This group, whose name means 'hairy legs', are the barnacles (Chapter 6), the most common of all the fouling animals, that is those found on ships' hulls, buoys, pier piles and so on (Chapter 12, Colour Plates 36, 37 and 38). The barnacle nauplius has two horns — distinguishing it from other nauplii — and grows through six moults before changing into a **cyprid** — the advanced stage (Figure 2.11) — which looks like a tiny mussel with several legs and two antennae. After a short period in the plankton, when the cyprid does not feeds, it must find a place to settle: a rock, a ship's hull or a piece of driftwood. Once a suitable place has been found, the barnacle 'walks' on its antennae, testing the surface: then, if satisfied, it flips over onto its back, cements itself into position, and moults into a young adult (Figure 6.3).

Stomatopoda

The mantis shrimps are mostly tropical, and live in burrows or in rock and coral crevices. The eggs are carried by the female in a brood before hatching into advanced young that are among the strangest animals in the microscopic world. They have a cape-like cover, eyes on the ends of stalks, and several different types of legs: those at the front have long hairs; one pair have sharp penknife-like claws for catching food; and there are flat swimming legs towards the tail.

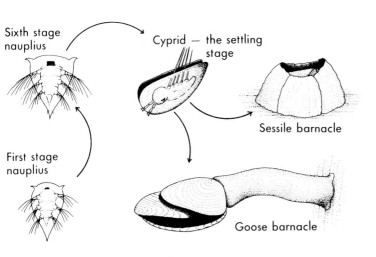

Sixth stage
nauplius

Cyprid — the settling
stage

Sessile barnacle

First stage
nauplius

Goose barnacle

Crabs, lobsters and shrimps are found world-wide and they all produce planktonic young. There is no planktonic nauplius, and the egg hatches into a well-developed young called a **zoea** which grows through several moults until it develops into an advanced stage called a **megalopa** (Figure 2.12). The head of the zoea has at least two horns, one projecting forward, the other pointing aft; there may also be a pair of horns projecting from each side. They have two large black or dark blue eyes, two or three pairs of legs, and a long tail ending in spikes. The advanced stage, or megalopa, is more complicated with additional limbs and coloured pigments developing. With detailed references it is possible to identify the young of different species, but at sea usually just the zoea and megalopa stages can be distinguished.

Decapoda

Figure 2.12: The young planktonic stages of a crab

Young zoea
stage

Older zoea
stage

Megalopa stage —
immediately prior to
adult

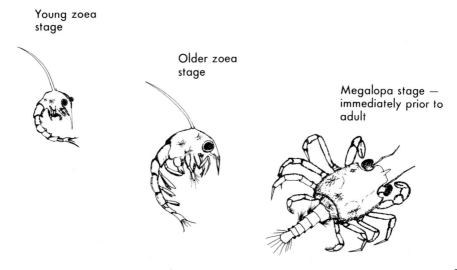

Plate 5. A large shrimp-like amphipod

Two groups of shrimp species are planktonic all their lives. They are over ten centimetres long, and have five pairs of legs and several paddle-like swimming limbs. The first three pairs of legs have small claws at the tips used in feeding. They have long antennae, eyes on stalks, and a point between the eyes. One group, *Lucifer* spp., occurs in all tropical and subtropical waters. They have a long, thin, flattened body with either black or red eyes. *Sergestes* spp., the second group, are among the several types of bright red, deep water prawns, that are rarely seen.

The remaining groups of marine crustaceans include animals that live their whole lives drifting throughout the oceans and at all depths.

Cladocera Water fleas range from one to five millimetres (a quarter of an inch) long and most live in fresh water though a few are marine (Figure 2.13). The freshwater flea *Daphnia* is used as aquarium fish food. Water fleas are almost entirely enclosed in a case like a shell. The head projects out of the case and has one black eye and two hairy antennae. These antennae are branched and give the group its name — the Greek *klados* meaning 'branch'. The eggs are held in a pouch in the shell and hatch as miniature adults.

Ostracods Mussel shrimps are rarely more than a few millimetres long. The body is completely enclosed in an eliptical shell-like case so that they resemble tiny mussels (*ostrakon* is Greek for shell). In some species the case has sculptured patterns and may be coloured grey, brown or green and occasionally yellow or red. The head has two antennae with long bristles for swimming (Figure 2.13). Eggs hatch into the typical nauplius stage but the young are held in the case until they develop into tiny adults, at which point they are released. Mussel shrimps were the first crustaceans in which bioluminescence was observed (Chapter 3).

Copepods Named after the Greek word meaning 'oar-footed', copepods are

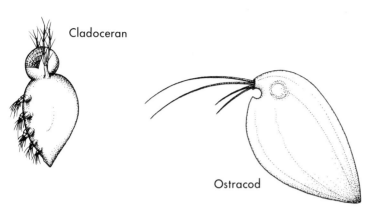

Figure 2.13: A cladoceran, the water flea, and an ostracod, the mussel shrimp

probably the most important link in the marine food chain: they are a major part of the diet of many animals ranging in size from whales to fish. Almost 70 per cent of zooplankton are copepods, and it is possible that they may be the most numerous group of animals on earth. They grow to between one and several millimetres long and are shaped like a grain of rice, with two antennae at one end and a tail at the other. Most have an eye; sometimes two.

Figure 2.14: Young and adult Copepods

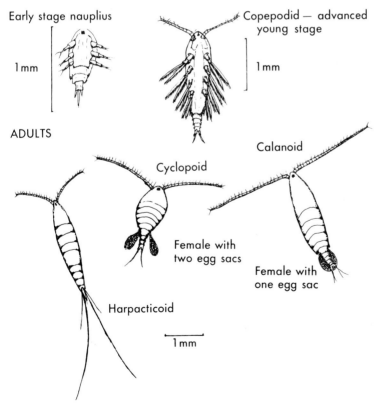

The majority of copepods feed on plant plankton although

some are carnivores and eat other zoo-plankton. On either side of the mouth are small hairy limbs rather like tiny arms; the hairs make an efficient net. A current of water passing through the net is created by the 'arms' beating and food particles are trapped and pushed into the mouth. The beating arms make the animals glide slowly through the water, but they can also swim with a jerky motion by beating the legs and laying the antennae alongside the body. At rest the antennae extend outwards to stop the animal from sinking too quickly. Some species are brightly coloured: red, orange, blue, purple or black. Many species are bioluminescent (Chapter 3).

There are males and females, the male sometimes having a jointed antenna and often an extra limb used for grasping the female during mating when he plants a specially made bag of sperm onto her genital opening; the eggs are fertilised inside her body. Some females shed the eggs directly into the water after fertilisation; others produce one or two sacs in which to carry the eggs, which are sometimes a brilliant translucent blue or green colour. The eggs hatch into nauplii, which moult five or six times before developing into **copepodids** — the advanced stage (Figure 2.14). The copepodid resembles the adult but has fewer segments, and it moults four times before the adult stage is reached. Nauplius to adult may take from a week to a year depending on the species.

Four types of copepod can be recognised (Figure 2.14):

Calanoids are the largest and most common copepods, with over 1,200 species. The body is distinct from the tail, and the antennae are usually long and occasionally articulated with a joint for holding the female in copulation. Some may also have a modified leg which serves as a clasper to plant the bag of sperm. The female carries one egg sac.

Cyclopoids, with over 1,000 species, are smaller than the calanoids and have shorter antennae. The body is rounded and distinct from the tail. The female carries two egg sacs.

Harpacticoids. There are about 1,200 species and the majority live on the sea-bed, but there are a number of planktonic types. The body is elongated and not distinct from the tail so they are shaped like an exclamation mark with two long bristles at the tail end.

Parasitic Copepods. Flying fish found on deck often have long thin worm-like animals attached. These are parasitic copepods. They attach to the gills, fins or skins of fish and are sometimes called 'fish lice'. Some other parasitic copepods live in sea squirts, shellfish and starfish.

Most parasites have several adaptations for their particular way of life, such as hooks for grasping the host and piercing mouth-parts for sucking body fluid. But the degree of modification varies, and some species are totally unrecognisable as belonging to any particular group. Usually it is the female that shows the greatest degree of modification: often the male and young stages are more typical of the group.

A typical life-cycle of a copepod parasite can be seen in the copepod that lives on the gills of the Atlantic Salmon. It is known as the salmon gill maggot. The advanced young (copepodid) stage attaches to the gills of the salmon as it enters coastal water. The male copepod matures quickly and copulates with the still immature female; then he dies. The female passes through the final moult, matures, and becomes firmly embedded in the fish's gills. She produces a sacful of fertilised eggs which may be over a centimetre long; sometimes several clutches are produced. The eggs hatch into a nauplius which develops into the copepodid and the cycle continues.

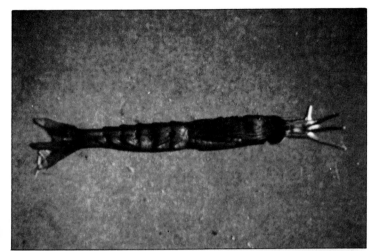

Plate 6: An opossum shrimp or mysid
(a) top view.

(b) side view — the balance organs can be seen in the tail as two translucent spots.

Opposum shrimps are so called because they carry their young in a pouch. They grow to between one and three centimetres long (just over one inch) although one giant reaches 35 centimetres (nearly 14 inches). Mysids have a pair of antennae, split into two, and a pair of eyes, each on the end of a stalk. The front part of the body, the thorax, is covered by a shield-like cape, and the legs are underneath (Black and White Plate 6). The first pair are used for holding food: the rest are hairy and, together with the flattened tail, are used for swimming. The adult female has several flat legs modified to make the brood pouch for holding the eggs. In the tail are two

Mysids

33

balance organs similar to the ones in jellyfish (page 23), but in this case the grain is made of calcium fluoride.

The 'cape', or **carapace** is attached only at the front end; hence one way to test whether the specimen is a mysid is to try and lift the back of the cape with a pin. If it is attached, then the specimen is not a mysid. The carapace is well supplied with blood vessels and works as a gill where oxygen dissolves from the water into the blood.

There is no nauplius stage: the eggs hatch and leave the brood pouch as well developed, but miniature, adults.

Cumacea   The cumacea are about ten millimetres long (under half an inch) and they live in burrows on the sea-bed but may be found at the surface after stormy weather or in breeding swarms at night. They look like small shrimps with the carapace projecting forward over the eyes and mouth, and a tail with bristles (Figure 2.15). The eggs hatch into an advanced stage of development, but with the last pair of legs not quite fully grown.

Isopods   Isopods are named from the Greek meaning 'equal legs', and are so called because there is no difference in the structure of their legs. All the animals in this group look like that most well-known of all isopods — the wood louse. Most species crawl on seaweeds or burrow into sand or wood. The wood borer (or gribble) *Limnoria* spp. is a commercial pest, attacking wooden boat hulls and pier piles (Chapter 12). Planktonic isopods are usually brown and between 5 and 14 millimetres long (just over half an inch). The body is in segments, each with a pair of flat legs, like paddles, for swimming (Figure 2.15). Underneath at the back are a pair of 'doors' that protect the gills. The eggs are held in a brood until they hatch at an advanced stage.

Figure 2.15: A cumacean (a) and an isopod (b)

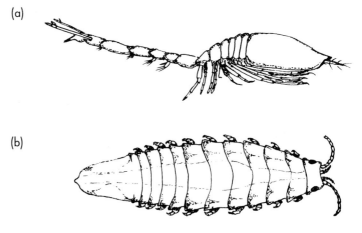

(a)

(b)

Several isopods are parasites and are either adapted to living in the gills of fish, or attached to their skin (Colour Plate 27). The mouth is used for piercing and sucking the body fluids from the fish. One species, *Gnathia*, is a fish parasite when it is young, but swims freely as an adult — not feeding, but surviving instead on food accumulated when it was parasitic. Other isopods have evolved to live as parasites inside crabs.

The name amphipod means 'two kinds of legs'. These 'shrimps' are flattened sideways and have two types of legs; one for walking, the other for swimming. In the planktonic species all the legs are used for swimming. They have large eyes, on the sides of the head (not on stalks) and there are two pairs of antennae. The females keep the eggs in a brood chamber under the abdomen until they hatch into an advanced stage, from which they develop directly into young adults.

Those normally caught at sea are called **hyperiid** amphipods (Figure 2.16). One species, *Hyperia galba*, has an interesting relationship with the lion's mane jellyfish (Chapter 4 and Colour Plate 2). Several *Hyperia* live on the jellyfish without coming to any harm and even take bits of food from their unusual 'host'. *Phronima* is another intriguing species. This animal feeds on planktonic sea squirts (Chapter 7 and Black and White Plate 9) which have a transparent body. The *Phronima* eats out the middle of the sea squirt leaving a barrrel in which it lives with its young. The animal has been reported pushing several barrels like a string of beads — each full of their young. Unfortunately it is difficult to catch *Phronomia* together with its 'barrel' house.

a. *Hyperia* sp.

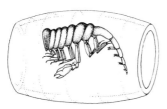

b. *Phronima* sp. in its 'glass house'

c. *Cyamus* sp. — whale louse

A few speces of amphipods are fish parasites with piercing mouthparts, feeding on the body fluids of their host. Another group — the **cyamids** — are commonly called whale lice (Figure 2.16). They have hooks on their legs for clinging to the whale but it is unlikely that they are true parasites as they do not harm the whale but feed instead on plankton and debris on the whale's skin.

These animals get their name from a word meaning 'shining light'. They are shrimp-like animals that produce very bright displays of bioluminescence. They resemble the mysids except that the 'cape' is attached both fore and aft and does not have blood vessels or work like a gill; neither are there any balance organs on the tail. The gills look like bunches of feathers attached to the tops of the legs (Colour Plate 28). The head has two antennae and a pair of eyes on stalks; the legs and flat tail are used for swimming.

The most distinguishing features of these animals are the rows of red spots on the lower parts of both sides. These are the light-producing organs or **photophores** (Chapter 3).

During mating the male plants a bag of sperm onto the female and the eggs are fertilised inside her body. They hatch into nauplius stages that will develop through several moults before becoming adults. They feed on particles filtered from the water by the hairs around the mouth and some, such as *Euphausia superba*, grow to six centimetres (two and a half inches) long. This species, known to the whalers as 'krill', provides food for the great whales and lives in large swarms, often covering an area of several square miles. Such a swarm may look like a huge constantly moving cloud in the water. It may occupy a layer five metres (15 feet) or more deep, but the surface water contains the greatest concentration which may reach 63,000 animals per cubic metre. All the surface animals are adults or near-adults because the young stages are restricted to between 700 and 1,500 metres (2,300 to 4,900 feet) deep where the water is warmer in the polar regions. A moderate-sized blue whale may consume a ton of krill at one feeding and make up to four such feedings a day!

The pycnogonids — sea spiders

Sea spiders live in all oceans though commonly in cooler areas. They live mainly on the sea-bed, crawling among seaweeds and animals, but some are able to swim using flapping movements of the legs and can sometimes be picked up in plankton samples. The animals are not true spiders but resemble the 'daddy-long-legs' without the wings. They have between four and six pairs of legs, and a thin body built in sections (Figure 2.17 and Black and White Plate 7). The mouth is at the end of a downward pointing snout and they attack sea anemones, sea firs and sponges.

The female lays the eggs and the male fertilises them and picks them up to carry them on his legs until they hatch. The young sea spider may stay attached to the legs or may live as a parasite on sea firs or corals. Either way they eventually moult into adult pycnogonids.

Figure 2.17: A mite (Acari) and a sea spider (Pycnogonida)

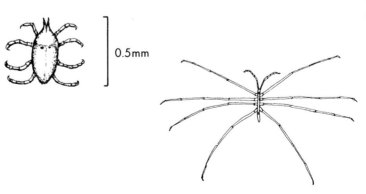

0.5mm

The acari — mites

There are very few marine planktonic mites because they are generally crawling animals, but some species live in the oceans by crawling over other animals. They have a flat, almost circular body up to about two millimetres long, with eyes on their backs and four pairs of legs (Figure 2.17). The females produce eggs that are fertilised by the male, and the young hatch with only three pairs of legs. After several moults they develop the full complement of legs and grow into adults.

Plate 7: A sea spider or pyconogonid

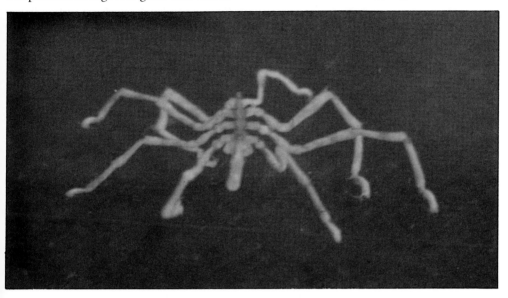

These animals live mainly on the sea-bed, but they start life as a small drifting barrel-shaped organism that has rows of hairs over its body (Figure 2.18). This young stage develops into a pyramid shape with long thin arms that carry the small growing adult. When it is fully developed, the carrier is shed and the young adult sinks to the sea-bed to start an independent life. Sometimes miniature starfish can be picked up in coastal plankton samples.

The echinoderms — starfish, sea urchins and sea cucumbers

Figure 2.18: The development of brittlestars, starfish and sea urchins (echinoderms)

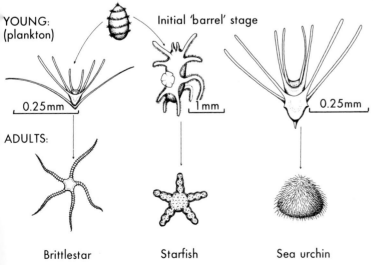

YOUNG: (plankton)

Initial 'barrel' stage

0.25mm

1mm

0.25mm

ADULTS:

Brittlestar

Starfish

Sea urchin

The bryozoa — sea mats

Sea mats grow encrusted on rocks, seaweeds, driftwood or ships' hulls. Some resemble pieces of coral, while others look like little plants, but when seen under a magnifying lens they appear like tiny boxes, each with an animal living inside. The young are planktonic and resemble a trochophore, but are more triangular, with a tuft of hairs on the apex and two rings of hairs around the base looking rather like a pair of spiked wheels (Figure 2.19). The animal and its internal organs can be seen inside the clear outer shell.

Figure 2.19: The development of sea mats (bryozoa)

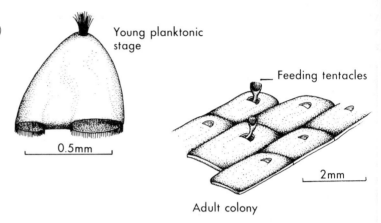

Young planktonic stage

Feeding tentacles

0.5mm

2mm

Adult colony

The chaetognaths — arrow worms

There are about 50 species of arrow worm. All except one are planktonic and they are found world-wide, though mostly in tropical waters. They are shaped like a feathered dart, and are usually about two centimetres long though some species reach ten centimetres. The main part of the body has fins on either side for stability and a fin around the tail for swimming. The animals are transparent except for the eyes, which look like tiny black dots, and the jaws, which are brown and look like a set of curved spines (Figure 2.20). They turn opaque when preserved.

Figure 2.20: Three species of arrow worms (chaetognaths) from the North-east Atlantic

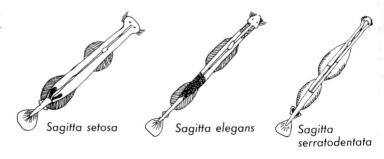

Sagitta setosa   *Sagitta elegans*   *Sagitta serratodentata*

When catching food the arrow worm darts forward and seizes the prey which is swallowed whole and can often be seen in the gut. They usually eat copepods, but occasionally prey on small fish and other arrow worms.

Arrow worms migrate vertically, swimming upwards using the tail fin and then sinking slowly with the side fins outstretched.

The animals are characteristic of certain types of water, hence by identifying the species it is possible to find out whether you are in coastal or oceanic water, tropical or temperate water. For instance, as you approach the British coast from the west the species changes; temperate Atlantic oceanic water contains the species *Sagitta elegans*, whereas coastal water has *Sagitta setosa*. Where you get both species then the water is mixing.

There is only one marine insect found at sea. This is *Halobates*, the ocean skater (Black and White Plate 8). It is found world-wide, in most tropical waters. It walks on the surface film of water like the freshwater pond skater commonly found on lakes and slow-moving rivers.

The insects

Plate 8: The ocean skater, Halobates sp.

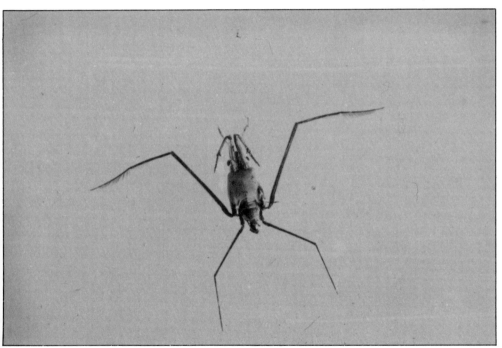

Chordates are animals with a stiff backbone or 'cord'. The planktonic marine chordates are: the sea squirts (**urochordates**); the lancelets (**cephalochordates**); and the fish (**vertebrates**). All have planktonic young but some sea squirts are also planktonic (Chapter 7).

The chordates — sea squirts and vertebrates

The sea squirts are enclosed in an outer skin or tunic, hence the animals are sometimes referred to as **tunicates**. Members of one group live attached to the sea-bed and produce planktonic young; another group are entirely planktonic (Black and White Plate 9). these have a clear or semi-transparent tunic with muscle bands around the animal so they resemble barrels. Water is sucked in through an opening at the front, passes through a basket network inside the animal that filters out small plankton for food, and is then pushed out of the rear of the animal thus creating a jet to push it through the water. Most sea squirts are

Urochordates — sea squirts

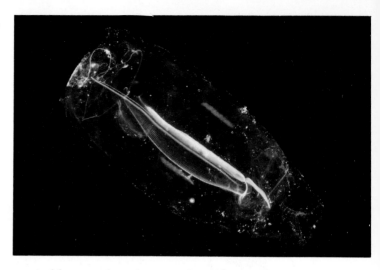

tropical but some species occur in cool temperate regions.

They often grow through a complicated life-cycle starting with a planktonic young stage which looks like a tadpole with a distinct head and a long thin tail (Figure 2.21). This develops into an individual with a long fleshy projection at one end which grows and adds buds of new sea squirts along its length. At this stage the tadpole young is called a 'nurse'. Eventually this nurse has a long string of attached sea squirts and at some time each one will break off to start an independent life in the plankton.

Figure 2.21: The life-cycle of a planktonic sea squirt

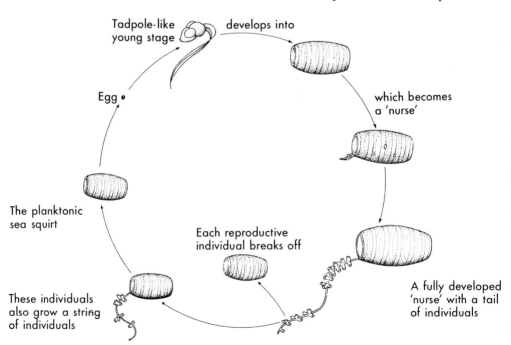

Some sea squirts live together in colonies with the individuals usually about ten millimetres long, although the colonies may reach up to two metres (Chapter 7 and Colour Plate 45). Some species are responsible for very dramatic displays of bioluminescence (Chapter 3).

All sea squirts produce a tadpole-like young stage, but in one group the typical form of the adult stage is missing and the tadpole stage develops the reproductive organs. These are called the **Larvacea**. They live in a delicate jelly-like 'house' which is also an intricate mechanism for catching food (Figure 2.22). Water is drawn in through a net that excludes large particles, and then passes through a fine net which catches the food — extremely small planktonic plants. At the bottom of the 'house' is a hatch through which the animal can escape if danger threatens. When the specimen is caught the house is invariably broken, but in good conditions the animal will build a new one in a short time.

Figure 2.22: A Larvacean in its 'glass house'

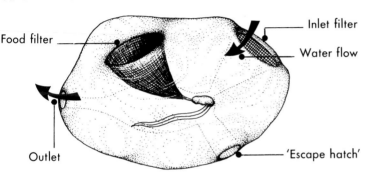

Food filter

Inlet filter

Water flow

Outlet

'Escape hatch'

It is difficult to distinguish between a young sea squirt and a larvacean, but usually the tadpole young are not found in plankton samples. It is possible that although planktonic offspring have been seen in laboratory animals, in the wild they may only occur near the sea-bed.

Cephalochordates — the lancelets

These animals resemble tiny eel-like fish. The adults live in sand or gravel near the shore but the young, which are practically transparent except for a pair of black eyes (Figure 2.23), migrate to the surface at night.

Figure 2.23: A lancelet, Branchiostoma spp.

0.5cm

Vertebrates — the fish

The tiny eggs and small fry of fish are to be found in most plankton samples. Their identification is extremely difficult; the colour of the embryo, yolk sac and eyes are all important. It is worthwhile making a note of these characteristics while the eggs are fresh, as they become obliterated in preserved specimens. When first caught most fish eggs are clear round balls with the curved embryo inside (Figure 2.24).

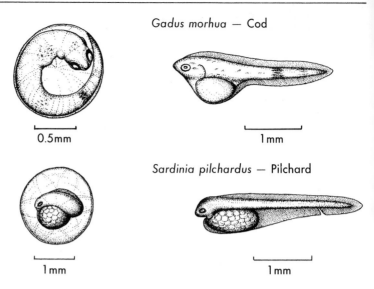

Figure 2.24: Fish eggs and fry

Gadus morhua — Cod

0.5mm

1mm

Sardinia pilchardus — Pilchard

1mm

1mm

Many fish shed their eggs and sperm directly into the water, but others stick them to seaweed and stones on the sea-bed. Rays, dogfish and some sharks (Chapter 8) produce eggs inside hard brown cases with stringy projections which fasten the cases to seaweeds. These are often found on the shore and are commonly called mermaids' purses.

Fish hatch as fry which may be totally different from the adult, hence even at this stage identification is difficult. Initially the fry carries a store of yolk for food but by the time the yolk is finished the mouth and anus have developed and the fish starts to feed on other planktonic animals. Through progressive stages the fish grows into an adult.

3. Bioluminescence
Living light

While sailing a little south of the Plata on one very dark night the sea presented a wonderful spectacle. There was a fresh breeze, and every part of the surface, which during the day is seen as foam, now glowed with a pale light. The vessel drove before her bows two billows of liquid phosphorus, and in her wake she was followed by a milky train. As far as the eye reached, the crest of every wave was bright, and the sky above the horizon, from the reflected glare of these vivid flames, was not so utterly obscure as over the vault of the heavens.

Charles Darwin, HMS *Beagle*
December 1833.

Plate 10: Bioluminescence in the Arabian Gulf caused by the bow wave of a ship

Bioluminescence, or phosphorescence, is probably the strangest phenomenon of the marine world (Black and White Plate 10). In the seventh century it was thought to be caused by the sea absorbing the sun by day and emitting it at night. Another view was that the flashes were caused by friction between the waves, or between the waves and an object such as the side of a ship. In the mid-eighteenth century, Benjamin Franklin thought that it was an electrical discharge between salt and water particles. But later in the same year two Venetian naturalists proved that bioluminescence in the Adriatic was caused by the single-celled animal *Noctiluca*. Sometime later another naturalist observed bioluminescence in the Mediterranean jellyfish *Pelagia noctiluca* — though this was only confirming data recorded by the Roman naturalist Pliny.

Today it is common knowledge that many organisms can produce light. On land there are fire-flies, glow-worms and even luminescent mushrooms; in the sea there are a large number of luminescent species (Table 3.1).

TYPES OF BIOLUMINESCENCE

It is difficult to classify bioluminescence because there are so many different types, and because observations vary. For instance, the apparent intensity will depend on how long the observer has been in the dark, and perception of colour differs greatly between individuals. The best way to record luminescence is to note precisely *what you see, when you see it*. Take note of the colour, size, extent, duration and anything else possible; perhaps then, in time, we may discover more about this phenomenon. Below are some experiments that will help to identify the type of luminescence:

1. What is the effect of shining a light on the area?
2. What is the effect of the echo sounder?
3. If you have stopped, what happens when the engine is started? And vice versa.
4. What happens if the radar is switched on or off?

Then take a sample of water, but keep it in the dark at first:
5. Is there any luminescence in the sample?
6. What happens if the sample is shaken?
7. Does the luminescence change in torchlight?

After making these observations, the sample can be taken into the light and using the notes from Chapter 2 it should be possible to find out what type of animals are in the water. But if you want to see the luminescence again, the sample should first be left in the dark for at least 20 minutes.

There are many types of bioluminescence and it is up to the observer to state what is seen. However, many of the luminescent phenomena will fit into one or several of the following categories.

MILKY SEA OR WHITE WATER

This effect is said to give an area a ghostly appearance. It is an even diffuse glow in the surface waters, or immediately beneath, that extends over a large distance. There are no distinct shapes and the glow is not affected by mechanical disturbances. Milky sea is commonly recorded between July and September in the Arabian Sea and Malacca Straits (Map 3) but it has been recorded elsewhere and at other times of the year. There is normally a definite edge to the area although the wind may blow the luminescence into slicks or streaks. Some seafarers have noted a change in the seawater temperature which indicates that the vessel has moved from one body of water to another. A 'fishy' smell has also been recorded, which could be due to a large amount of dead plankton. In a true milky sea there is no luminescence in a sample of water, nor can any be induced by shaking or flashing a light.

The cause of a milky sea is not definitely known because samples have yet to be examined in a laboratory. A possible cause could be waxes from the mass death of copepods and the breaking up of their bodies. But bacteria (small organisms which can only be seen through a high-power microscope) are the only organisms known that can luminesce continuously. The Arabian Sea is often very rich in these in the summer months because of upwelling water caused by the monsoons, and there is an increase in the amount of algae on the surface. This algae is not luminescent itself but it could be a base on which the bacteria can grow.

LUMINESCENT WHEELS

Luminescent wheels are waves of luminescence travelling outwards very quickly in an expanding circle from a central hub which may also be luminescent. The formation varies from two metres to one and a half kilometres in diameter, and sometimes 'spokes' can be seen radiating from the hub to the rim. Some patterns revolve — this is a true luminescent wheel — and the direction of rotation may remain constant or it may change repeatedly. Wheels may start with an eruption of luminescence from deeper water which 'explodes' on the surface and spreads out in a large circle. At other times there may be many wheels, interlocking or separate. Wheels are mainly found in shallower water and almost exclusively in the Indian Ocean and South China Sea (Map 4).

Luminescent wheels are perhaps the most puzzling of all bioluminescence observations. Some people have tried to explain them in terms of the ship's wake and waves, but this is not entirely satisfactory. Records from hydrographic survey ships have reported bands of light resulting from using an air-gun for taking seismic profiles of the sea-bed — bioluminescent flashes coinciding with the firing of the gun. It is possible that this phenomenon is due to seismic disturbance on the sea-bed stimulating large numbers of luminescent plankton. But there have been no direct measurements of seismic activity or of the luminescent plankton in the relevant areas, nor has it been proved that seismic waves can stimulate organisms to luminesce.

LUMINESCENT BANDS

This effect, whose distribution is shown on Map 5, may be associated with a large luminescent wheel; it is sometimes difficult to distinguish between the spokes of a large wheel and a system of bands. The ship may appear to be sailing over a series of parallel lines that may be moving. Bands have been recorded passing at rates of between two and five per second. The lines are usually long — from 100 to 400 metres, or even stretching to the horizon — and they vary from 60 metres wide and 300 metres apart to one and a half metres wide and one and a half metres apart. They have distinct edges which distinguishes them from streaks caused by winds or currents, and from trails due to flying fish or dolphins. Streaks and trails are shorter, less structured and not as persistent. Once again a possible explanation for luminescent bands is seismic activity, but the evidence is not conclusive.

ERUPTING LUMINESCENCE

This phenomenon (Map 6) is also often associated with wheels. Descriptions from observers are often dramatic — luminescent balls of light that erupt from the depths and explode on the surface. If a wheel is not formed then the luminescence usually fades. The size of the ball varies between 30 centimetres (one foot) and two or three metres (up to ten feet) in diameter before spreading into discs of 30 metres (100 feet) diameter. Often 'blobs' of luminescence rise into the ship's wake but these are probably caused by the propeller churning up some unlucky animal.

Erupting luminescence characteristically occurs in deeper water and once again is possibly caused by submarine seismic disturbances.

PATCHES AND SPECKLES

These are probably the most common types of bioluminescence. They are small pinpricks of light in the water, seen from the ship or in a sample anywhere in the world. They are caused by tiny planktonic organisms disturbed by the ship, and their colour, size and duration vary according to the species. Sometimes speckles can be seen among wheels, bands or eruptions.

MISCELLANEOUS

Other luminescences have been reported in response to a light being shone on the surface. These may be reflections from the eyes of fish — usually lantern fish — characterised by erratic darting lights not 'luminescing' without the aid of the lamp. Occasionally it has been reported that observers were able to 'write' on the water using a searchlight. This is rare, but some animals will luminesce in response to light (Table 3.1).

Other types of luminescence have been said to be caused by switching on the radar. This is difficult to explain and many more observations are required before coincidence can be ruled out.

THE BIOLUMINESCENT ORGANISMS

Table 3.1 shows the main groups of organisms that are bioluminescent. Some emit light of a specific colour, but this depends on the observer's interpretation of the colour. Noting the brightness of the light is also up to the individual's interpretation of 'bright' and depends on how much the eyes are adapted to the dark. An animal that produces dim luminescence will appear bright to an observer who has been in darkness; bright luminescence will appear dim to anyone who has just come out of a well lit room.

THE PRODUCTION OF LIGHT

Whether the luminescence is continuous or in flashes or pulses gives a good indication of which organisms are responsible. The stimulus required for luminescence varies but it is usually mechanical disturbance — the passing of a ship, possibly seismic activity, or shaking in a jar. Even the organisms colliding with one another or the sides of a container may cause luminescence, and this may look as though no stimulus has been applied. Some animals will respond to light but most are inhibited. However, in all cases the bioluminescence is caused by a chemical reaction although the chemicals used may differ from species to species. In many small organisms the reaction takes place inside the body, and the whole animal will light up; in others the chemicals are squeezed into the surrounding water to react and so produce a luminescent 'cloud'. Many larger animals

Table 3.1: The bioluminescent organisms

Organisms	Luminescence type	Time period	Stimulus required	How light is produced	Distribution and notes
Bacteria (p. 44)	Milky sea	Continuous	Not known	Chemical	Arabian Sea, Malacca Straits. July — September
Dinoflagellates (p. 19)	Flash or sparkle	1/10 second	Mechanical	Chemical	World-wide. Inhibited by light. Extensive displays caused by large numbers
Radiolaria (p. 22)	Dim glow	1–3 seconds	Mechanical	Not known	May be responsible for many displays — but little is known.
Coelenterates:					
a) jellyfish stage (p. 22)	Entire animal glows	3/10 to 6 seconds	Mechanical	Chemical	Probably world-wide, little known.
b) siphonophores (p. 53)	Medium light	1–11 seconds	Mechanical	Not known	Probably world-wide, inhibited by light.
c) scyphozoa (true jellyfish p. 56)	Bright slime		Mechanical	Produced in cells on the surface	Only one known species: *Pelagia noctiluca*
Ctenophores (p. 23)	Bright pulse	2 seconds	Mechanical	Light passes down the comb rows	World-wide. Even the eggs are luminescent.
Annelids (p. 24 and 26)	Not known	Not known	Not known	*Tomopteris* (p. 26) has photophores	Atlantic. Some ragworms luminesce in breeding swarms.
Molluscs:					
a) Gastropods (p. 27 and 62)					Very little is known.
b) Cephalopods (p. 27 and 67)	Bright beams	Various	Mechanical some continuous	Photophores and luminescent bacteria	Most deep water but swarm to the surface at night. One squid has luminescent 'ink'.
Crustacea:					
a) Ostracods (p. 30)	Bright cloud	Not known	Not known	Chemical	Variable but especially east coast India, Malabar and Maldives. Bright displays may be caused by large numbers.
b) Copepods (p. 30)	External	2–37 seconds	Mechanical	Chemical	Distribution world-wide but little is known.
c) Euphausiids (p. 35)	Flash and beams	Perhaps up to a few minutes	Mechanical and breeding	Photophores	Open oceans. Photophores can rotate.
d) Decapods (p. 29 and 74)	Beams and clouds	1–5 seconds	Not known	Photophores and chemicals	World-wide.
Chordates:					
a) Urochordates (p. 39 and 78)	Bright may be coloured	5–30 seconds	Mechanical and light	Chemical	Mostly tropical. *Pyrosoma* (p. 79) well known. Other species also luminesce.
b) Vertebrates (p. 42 and 81)	Not bright	Various	Various	Photophores and luminescent bacteria	Mostly deep water but swarms to surface at night. Photophores point down.

have special luminescent organs called **photophores** where the reaction takes place (Figure 3.1).

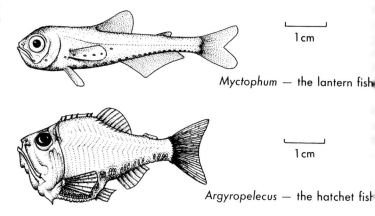

Figure 3.1: Bioluminescent fish

Myctophum — the lantern fish

Argyropelecus — the hatchet fish

Photophores have a clear window-like cover, a lens and a reflective layer (Figure 3.2) and are usually in the skin although some transparent animals have photophores inside the body.

Figure 3.2: Diagram of a photophore cut in half

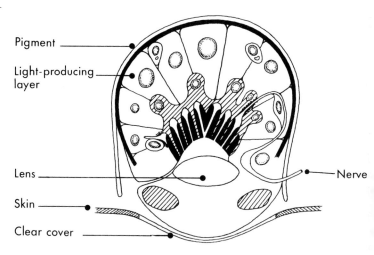

Pigment

Light-producing layer

Lens

Nerve

Skin

Clear cover

Many fish and squid have luminescent bacteria living in special hollows on their skin (Colour Plate 46). But bacteria luminesce continuously, which is not usually necessary or desirable, hence these animals have developed a shutter like an eyelid which closes over the luminescent area when the light is not required.

Bioluminescence is a world-wide phenomenon but dramatic displays are mostly restricted to tropical and subtropical waters. It is only seen at night as many bioluminescent animals — fish, squid, krill, mussel shrimps, copepods — are found in deeper water during the day but migrate to the surface at night, and bioluminescence is not bright enough to be seen in daylight: also some luminescence is inhibited by light.

THE REASONS WHY

There are so many animals that can produce luminescence that it must be beneficial to these animals. Most bioluminescence is blue-green (very rarely red and yellow may be seen); blue-green light penetrates farthest in seawater and most of the animals have eyes, so we know the light can be detected. Nevertheless it is unclear what function bioluminescence serves. There are probably a number of explanations.

A searchlight with which to look for food. In the dark some form of light would be an advantage in finding food. Euphausiids can rotate their photophores so that they point forwards; lantern fish *Myctophum* (Colour Plate 00) have been seen feeding on krill at night, perhaps using their photophores to find the food.

A lure. Some deep sea fish have lights that attract food. The angler fishes have a light on the end of an extension like a fishing rod, which dangles in front of their mouths; other fish have photophores on the inside of their mouths or cheeks to attract small inquisitive animals. A predator may also use a flash of light to confuse the prey that it is about to attack.

Separation. It has been suggested that extensive displays of surface luminescence may prevent the animals that would feed on plankton from migrating to the surface. The problems with this theory are that the luminescence must be long-lasting, which is not common, and that most luminescence requires a mechanical stimulus to work.

A warning. A flash may be a warning to predators not to eat a particular animal because it is poisonous or in some way 'nasty'. The midshipman fish *Porichthys* is luminescent and has a poisonous spine; jellyfish have stinging cells; sea gooseberries may be unpleasant to eat, and so on. Some species may imitate this type of luminescence even if they are not distasteful, and so avoid being eaten because the predator associates a particular kind of luminescence with an unpleasant organism. A flashing light may also be a warning to others that there is a predator nearby. Such a warning may not help the individual organism that is being eaten, but it may benefit the whole species.

A burglar alarm. A predator's enemies may be alerted by a flashing light. For instance a large fish could be attracted to a flashing animal that is being eaten by a smaller fish. This smaller fish may then be food for a larger fish. Once again this is only an advantage to the shoal of small animals — not to the one that is eaten.

An escape. A flash of luminescence may startle a predator so that the prey can swim off in the dark. Some lantern fish have a light organ on their tails that flashes when they are being chased, and copepods squirt out a cloud of luminescence before darting away if they are disturbed. This may be effective in temporarily blinding a pursuer. One species of deep sea squid has a luminescent 'ink' which it pumps out to confuse an attacker. In the opposite way some fish switch off their photophores when they are attacked which again leaves their attackers 'blinded'.

Camouflage. It may seem odd that flashing light could be camouflage, but many animals have the photophores directed downwards. Predators below will not be able to distinguish the fish from the surrounding light coming down from the surface whereas a dark shadow would stand out (Figure 3.3). Skin pigments, reflective surfaces, transparency, and behaviour such as vertical migration, may all help to camouflage animals.

Figure 3.3: How photophores can help in camouflage. In the semi-darkness, the shadow is easier to make out. Look at these drawings through half-closed eyes to see the effect.

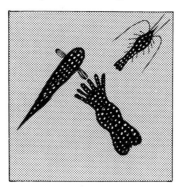 

Shoaling. Many animals live in shoals, and in the darkness bioluminescence may be a way of recognising the other members of a shoal. Some species can be distinguished by the arrangement of their photophores and perhaps animals can also recognise these distinctions. It is also possible that in a world of darkness light is the only way of telling other animals to keep out of a territory, therefore flashing lights may mark territorial boundaries.

Sexual displays. Recognising other members of the species, and sexually mature adults of the opposite sex, is important in the mating season. Some males and females have a different arrangement of photophores; in one species of fish only the males have photophores, and sexually mature females are attracted to the lights hence males may mate with more than one female. Many animals will increase their luminescence during the mating period; for instance, euphausiids normally respond to only mechanical stimulus but will respond to other light flashes during the breeding season, and some ragworms luminesce during their breeding swarms.

It is possible that some animals may carry out a luminescent courtship display. Other types of courtship behaviour are found in some freshwater fish so perhaps a similar display but involving light may help deepwater fish living in an area where recognising sexually mature individuals could be difficult.

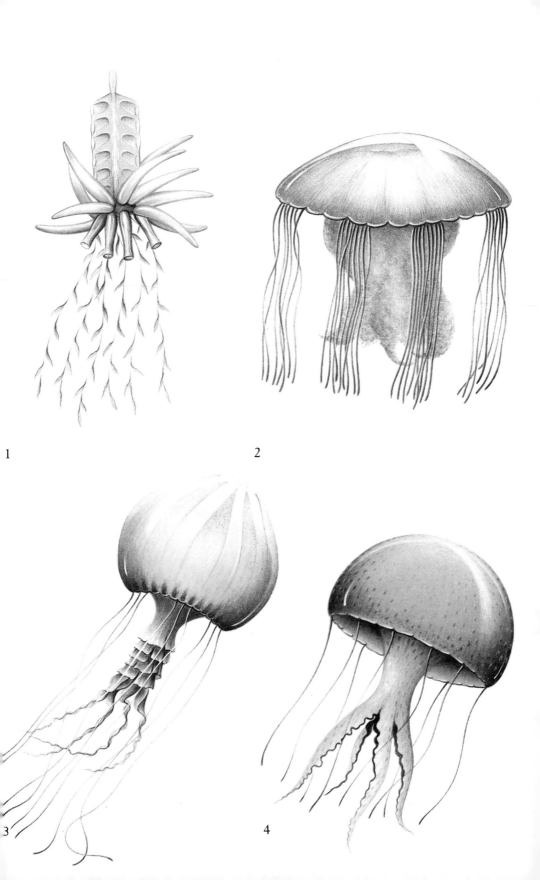

1

2

3

4

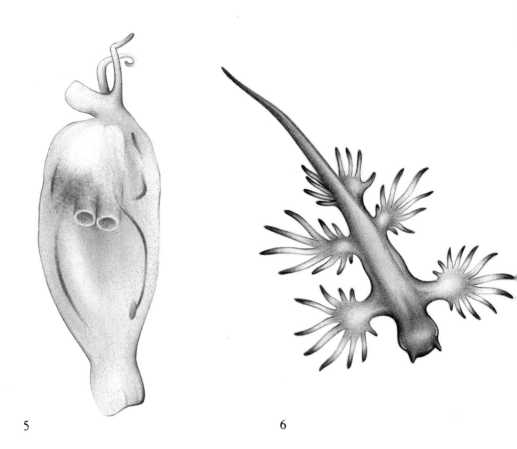

5

6

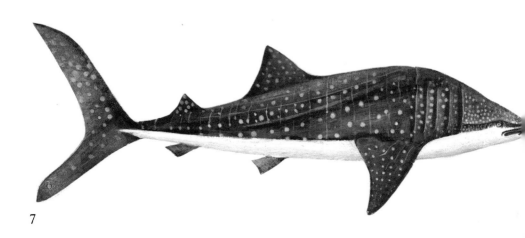

7

8

9

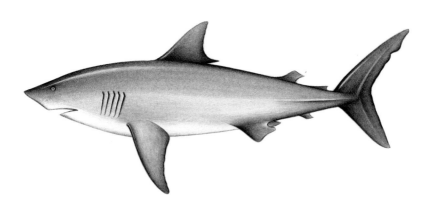

10

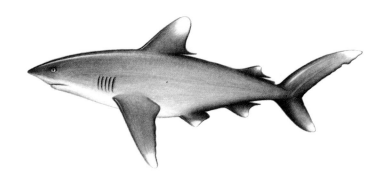

11

12

13

14

15

a

b

16

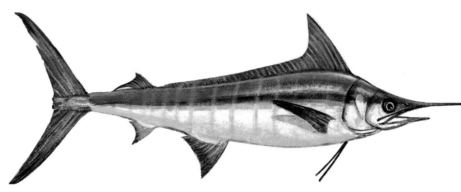

17

18

19

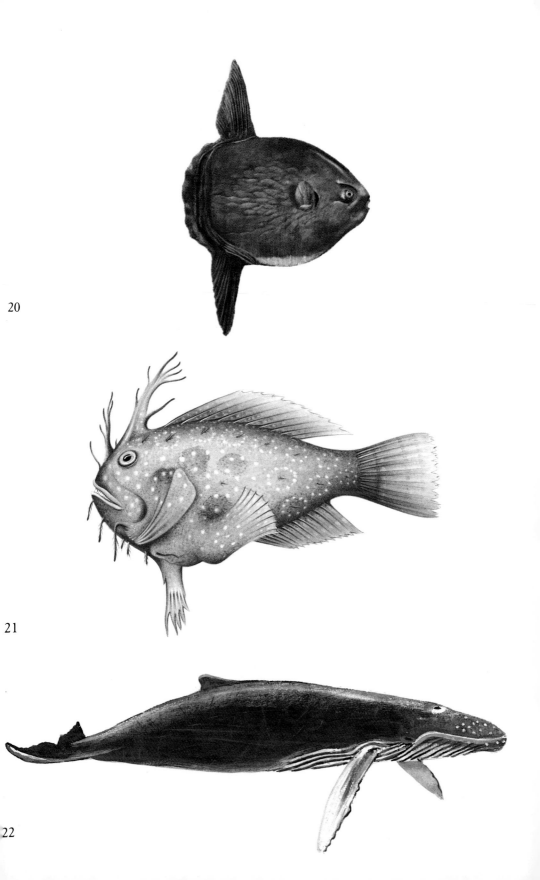

20

21

22

23

24

25

4. The Coelenterates
Jellyfish, Sea Anemones and Corals

Throughout the world — in port, at anchor, or sailing in the open ocean — jellyfish may be seen; some as big as footballs, others the size of pennies. Sea anemones decorate pier piles (Colour Plate 62), harbour walls, and shores, and coral gardens are found in warm, clear tropical water.

Despite this great variety all the animals in this group have similar features: the body is like a sack with one opening — the mouth — surrounded by tentacles, and they have an inner and an outer skin separated by a jelly-like substance (Figure 4.1). The tentacles are armed with stinging cells and are used for catching food. Each cell is like a fluid-filled balloon with a long pointed neck that is coiled up and pushed inside (Figure 4.1). When an animal brushes against the trigger, the neck is forced out and either pierces or wraps around the prey. Those that pierce are barbed, and usually inject a poison so there can be no escape. Once the cell has discharged it is useless, but it wll be replaced by another one in about 24 hours.

In some coelenterates the mouth points upwards. These have a thin jelly layer, sometimes an outer skeleton, and most are fixed to the sea-bed. These are the **polyp** type and include sea anemones and corals. Other coelenterates have the mouth pointing downwards, and a thick jelly layer — hence their common name — jellyfish. These are the **medusa** type.

Figure 4.1: The two types of coelenterate, and the coelenterate stinging cell

POLYP — e.g. anemone   MEDUSA — jellyfish

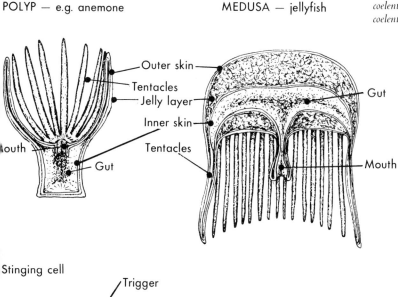

Outer skin
Tentacles
Jelly layer
Inner skin
Gut
Mouth
Tentacles
Mouth
Gut

Stinging cell

Trigger
'Door'

Undischarged   Discharged

CLASSIFICATION

Many coelenterates have a medusa stage *and* a polyp stage at some time in their life cycle (Figure 2.7), and their classification depends how long they live as either a polyp or medusa.

Figure 4.2: Classification of the coelenterates

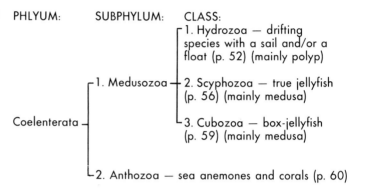

Jellyfish are the most commonly seen coelenterates as they are free-swimming: the polyps are mainly fixed to the sea-bed, although there are exceptions.

SUBPHYLUM 1. MEDUSOZOA

Class 1:
The Hydrozoa

Most of these animals grow attached to the sea-bed, weeds or rocks, and they are very common fouling animals (Chapter 12). They look like white whiskery or furry growths and are sometimes called sea firs or white weed. But there are two groups of animals in this class that spend their lives swimming or drifting on the oceans. One group comprises the **velellids**. These animals have a raft with a sail, and they drift along the surface pushed by the wind and currents. The tentacles hang from beneath the raft around the central mouth. Two species are found in this group.

Velella velella,
or by-the-wind-sailor

The species has a blue, oval raft up to ten centimetres long, with concentric rings; the sail is translucent and set at an angle across the top (Colour Plate 29). They are found world-wide (Map 7) though usually in the warmer oceans between 40°N and 30°S. Occasionally they occur farther north along the western margins of Europe and North America.

The short stubby tentacles catch the food — mainly copepods and other small planktonic animals. Male and female medusa stages grow between the tentacles, and when sexually mature these drift off and release eggs and sperm into the water where fertilisation takes place. The egg hatches into the young stage which sinks to deeper water where the currents often run in the opposite direction from those on the surface; hence this behaviour could be a way of spreading the species. Once the young has developed a float, the animal rises to the surface to begin its adult life. Gas is trapped within the concentric rings of the raft to make it buoyant.

By-the-wind-sailors are eaten by the purple bubble-raft snail

Janthina (Chapter 5) which also spends its life drifting. The snail slowly devours all the soft parts and leaves behind a ghost-like clear white husk.

Porpita has a circular raft up to eight centimetres in diameter, with club-shaped tentacles underneath (Colour Plate 00). This species is not commonly seen, probably because it is small, but they often occur in shoals in the Mediterranean and eastern Atlantic and in the tropical oceans.

Porpita porpita, no common name

    This species is similar to *Velella*; it feeds on similar food and is also eaten by the purple bubble-raft snail.

    The second group of animals in this class comprise the **siphonophores** (from the Greek *sipho* meaning siphon or tube). One species has a float that sticks above the surface; others have a float and swimming bells that create jets of water to propel the animal; a third type have only swimming bells.

    The animals are not simple, in fact every specimen is a collection of individuals, each doing a different job. It is a **colony**. Some individuals are for feeding; these have a mouth leading to one large stomach. Others catch food, and these have one long tentacle. A third type are for reproduction, and they produce the eggs and sperm. The colony develops through a process of budding. The egg hatches into a polyp that forms the float, all the individuals grow from buds on this float until the mature colony has the full complement of feeding, catching and reproducing members.

    Siphonophores are found all over the world, but mostly in warm seas. Many are small and transparent, but others are larger and include perhaps the most notorious of coelenterates — the Portuguese man-of-war.

This is the most commonly recorded animal in this group (Map 8). Its sting is well known — rather like a cigarette burn — but it is not fatal unless perhaps encountered by a bather who may drown as a result of many stings. For this and other jellyfish stings meat tenderiser can be an effective first aid treatment if it is quickly rubbed into the affected area, but generally the sensation lasts for at least 24 hours.

Physalia physalis, the Portuguese man-of-war

    Shoals of young *Physalia* resemble patches of scum or froth on the water. These grow until they look like blue bubbles, which are sometimes called 'bluebottles'. The mature colony has a large gas-filled float rather like an inflated plastic bag. The float may be up to 30 centimetres long and has a crest which acts as a sail — the Portuguese man-of-war, like the by-the-wind-sailor, being blown along by the wind as well as carried by the currents. The float is usually blue shot with pink, the crest is also pink but the tentacles, which may extend to several metres in length, are dark blue and the rest of the colony is blue to purple (Colour Plate 31).

    In *Physalia* and *Velella* the sail is set at an angle, to the left or right of the body — the animals are either left- or right-handed. The sail is set at 45 degrees to the wind direction and the colony beneath acts like a drogue, particularly in *Physalia*. Figure 4.3 shows a left-handed animal being blown by a northerly wind. The crest is aligned north-west to south-east (wind on the port quarter), the bulk of the body trails to the north-north-east, and the animal travels towards the south-south-west.

Figure 4.3: The Portuguese man-of-war, tacking across the wind

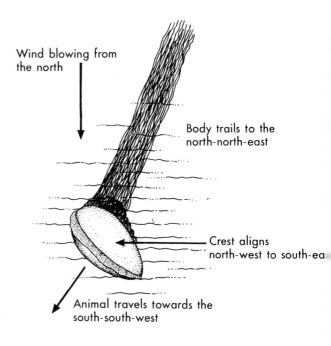

Wind blowing from the north

Body trails to the north-north-east

Crest aligns north-west to south-ea

Animal travels towards the south-south-west

It may be that having angled sails prevents the animals from crowding in one area because the wind tends to separate them. A right-handed animal will be blown south-south-east by a northerly wind, hence shoals usually contain mainly left-handed or right-handed animals. The northern hemisphere animals are mostly right-handed, southern *Physalia* are mostly left-handed. This means that the clockwise wind systems of the north Atlantic and Pacific oceans (Figure 1.2) disperses right-handed animals outwards so that they don't congregate in, for example, the Sargasso Sea in the Atlantic where the tentacles would become fouled amongst the *Sargassum* (gulfweed) and where food is in short supply. The anti-clockwise winds in the southern hemisphere disperse left-handed animals outwards. The separation of left- and right-handed animals between the two hemispheres is not a fixed rule, as variations in currents and wind direction cause both types to drift across the equator. More information is required on the distribution of the different types of *Physalia*.

Physalia has only a limited behaviour apart from catching food. The float does not remain upright but in calm sea the crest collapses, and rises again when stimulated by the wind. In a tank, the animal will dip each side alternately in the water to keep the float moist.

The Portuguese man-of-war feeds mainly on fish caught and paralysed by the tentacles and stinging cells. The food is then pulled up to the feeding parts which suck pieces into the stomach. However, despite the danger, one species of fish, *Nomeus gronovi* — the man-of-war fish (Chapter 8) lives unharmed among the battery of tentacles with their armoury of stinging cells. The fish must be either immune to the sting or be able to avoid the tentacles. However, if the fish is confined in a container with the *Physalia* it behaves as if it has been stung. The fish is camouflaged with blue stripes to blend in among the tentacles where it is protected from larger predators. While there it may also attract larger fish that become food for the Portuguese man-of-war. This remarkable relationship can be found in other coelenterates (Colour Plate 32) and certain species of fish.

Some species of purple bubble-raft snails, and turtles (Chapter 9) will eat *Physalia*, and certain types of shore crabs (Chapter 6) wait to feed on any that are stranded on beaches although the stinging cells remain active for a while after the colony has died — a fact worth remembering if you find a 'dead' Portuguese man-of-war.

Other siphonophores are not such passive drifters as *Physalia* because they have pulsating swimming bells. They are not usually found on the surface but just below, and hence are more likely to be seen when sailing slowly or becalmed.

Physophora has a sausage-shaped float above two rows of swimming bells, and pink finger-like tentacles which have powerful stinging cells. The animals are very delicate, resembling exotic water lilies (Colour Plate 1) and usually break up in a tow net. To observe the whole animal it must be gently lifted out of the sea in a bucket. *Physophora* is found world-wide in warm tropical waters.

Physophora hydrostatica

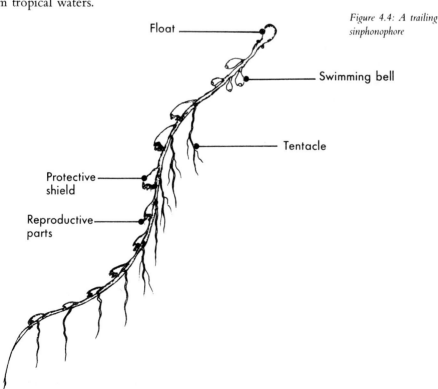

Figure 4.4: *A trailing sinphonophore*

Float

Swimming bell

Tentacle

Protective shield

Reproductive parts

Other species in this group have a float with a tail which has the individual members of the colony in clusters along its length. Shield-like projections protect the reproductive parts. Figure 4.4 shows such a colony. The species *Apolemia*, which occurs throughout the Mediterranean area, has a tail which may reach a length of 20 metres (66 feet) or more.

The third type of siphonophore has one or two swimming bells towing a trail of individuals like a train pulling carriages, although usually only the bell or part of the colony is picked up in samples. The following is an account of how the specimen in Black and White Plate 11 was caught in the Indian Ocean.

Plate 11: A siphonophore that was caught in a bucket in the Indian Ocean.

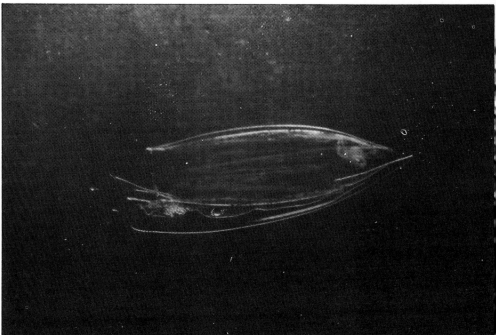

'There were bioluminescent flashes in the wake, and using a bucket on a rope a sample was collected from the aft deck. However, in a lighted room nothing could be seen in the water until it was poured through a fine net; even then the transparent pulsating animals were difficult to find.'

Sometimes square or rectangular pieces of 'jelly' may be picked up on plankton samples. These are the buoyant protective parts of a siphonophore, which may continue to pulsate for a short while.

Class 1.
The Scyphozoa —
true jellyfish

There are approximately 250 species of jellyfish, distributed all over the world. It is impossible to include all those that may be seen, but by using these notes some of the common ones can be identified.

Jellyfish are medusae for most of their lives; some in fact no longer have a settled polyp stage. There are flat species, shaped like saucers, others are more rounded, like a bell; but they al have a thick layer of jelly, and a mouth in the centre of the lower surface, usually with four or eight 'lips' like long flaps or frilly arms (called **oral arms**). In most jellyfish the tentacles hang from the outer edge of the bell, but in one group — the **rhizostomes** — the tentacles are clumped together in the centre. At intervals around the edge there are balance organs and light-sensitive cells (Colour Plate 26).

Large fish, turtles, and in some areas man, eat jellyfish and the stranded animals are devoured by scavenging shore crabs and gulls.

The aptly-named moon jelly is a saucer-shaped jellyfish that grows to a diameter of 50 centimetres (20 inches) and can be found in coastal waters all over the world. It is usually pink or purple, with several fine radiating lines which are tubes, some carrying food from the stomach to the rest of the body, others taking waste material back to be discharged through the mouth. Around the margin are many fine short tentacles that can be seen only when the animal is viewed close up. A distinctive feature of the moon jelly are the four three-quarter round organs looking like two crossed figures of eight which can be seen, even at a distance, clustered around the centre. These are the reproductive organs: ovaries in the female; testes in the male. The mouth has four long oral arms which can reach out to the margin.

Aurelia does not have powerful stinging cells; furthermore the tentacles are not used to capture food. The animal eats small plankton — copepods, young barnacles, crabs and worms — which are caught like flies on fly-paper, in a mucus layer all over the body. This mucus is a constantly moving conveyor belt carrying the trapped animals to the edge of the disc where they collect in a 'blob' until picked up by one of the oral arms as it 'licks' the edge.

In the breeding season, male *Aurelia* shed sperm into the sea. These are attracted to the female and swim in through the mouth so that fertilisation is internal. The eggs develop into young that swim out of the mouth to be part of the plankton for a short while before settling on the sea-floor as tiny polyps (Figure 4.5). The following spring the polyps grow buds of new jellyfish stacked on top of each other like dinner plates; eventually each one detaches to start an independent life.

Aurelia aurita, the moon jelly

The two species in this family, *Cyanea capillata* and *Cyanea lamarcki*, are similar except for their size and colour. *C. capillata* is the largest jellyfish known; one specimen caught in Massachusetts Bay, USA, measured two metres (six and a half feet) in diameter. This species is dark red/brown with long tentacles and powerful stinging cells. *C. lamarcki* is smaller, rarely more than 30 centimetres (twelve inches) in diameter. It is deep blue to purple and the sting, though painful, is not as powerful. The following notes apply to both species.

The disc is flat but contracts to an inverted basin-shape when swimming. Surrounding the mouth are four thick wavy oral arms, rather like curtains, and around the edge are numerous fine tentacles, four or five times longer than the diameter of the body (Colour Plate 2). These form a fine net to catch food —

Family: Cyaneids, the lion's mane jellyfish

usually small fish and other plankton. Ropes which get covered with the broken pieces of the tentacles can inflict a painful sting.

Both species often have the company of a small crustacean, *Hyperia galba* (Figure 2.16), which lives on the jellyfish, taking bits of food. Another close 'friend' may be a young blue whiting or other fish of the cod family, often seen sheltering among the tentacles. In a time of danger the young fish can swim up inside the bell of the jellyfish without coming to any harm.

Figure 4.5: The life-cycle of Aurelia — the moon jelly

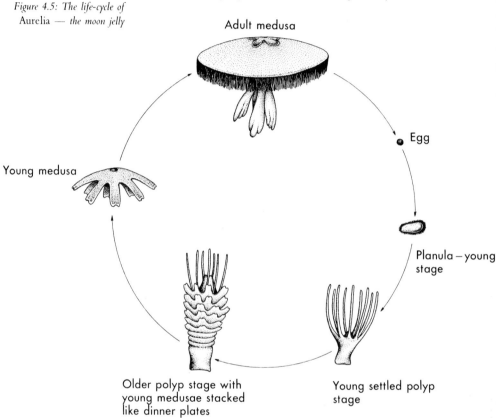

Adult medusa

Egg

Young medusa

Planula – young stage

Older polyp stage with young medusae stacked like dinner plates

Young settled polyp stage

Chrysaora hysoscella, the compass jellyfish

This species is bell-shaped, with distinctive brown lines on the top, radiating from the centre — hence its common name (Colour Plates 3 and 32). It grows to a diameter of 30 centimetres (twelve inches) and is opaque white to yellow with 24 tentacles up to 60 centimetres (two feet) long, and four thick arms about the same length around the mouth.

Compass jellyfish feed on other small coelenterates, arrow worms, sea gooseberries and sometimes small fish. They live in the surface waters of most temperate areas including the Mediterranean Sea.

Pelagia noctiluca

Pelagia is a pink, spotted mushroom-shaped jellyfish about ten centimetres (four inches) in diameter, with four oral arms the same length as the eight tentacles around the edge (Colour Plate 4). The spots on the bell are warts of stinging cells.

Pelagia produces a luminescent mucus which appears like luminous paint when it gets on your hands (Chapter 3, Table 3.1). When demonstrating this with a specimen caught while at anchor near Port Said, it was noticed that its sting is not powerful but like a series of pin-pricks which last for about 30 seconds. *Pelagia* is usually found in the surface waters of the North Atlantic, Mediterranean and North Pacific.

The Rhizostomes

This group of jellyfish do not have tentacles so they are easy to recognise. The bell may be up to 80 centimetres (32 inches) in diameter and is firm and rounded. In young specimens the mouth is surrounded by four lobes which divide into eight arms, but as the animal grows the arms blend together and obscure the mouth. However, they develop thousands of little sponge-like openings until the structure looks like a sprig of cauliflower. The stringing cells are on the arms, and the food — small planktonic animals — is sucked into the openings and through a system of canals until it enters the mouth and then the stomach.

Some fish are associated with certain species of rhizostome jellyfish. For instance *Rhizostomo pulmo*, which can grow to a metre (three feet) in diameter, is sometimes associated with young horse mackerel, *Trachurus*. In the port of Limon, Costa Rica, the ship was surrounded by a shoal of rhizostomes. They were a red and blue variety of the same species. Many had a small fish of the carangid family (horse mackerel, pilot fish family — Chapter 8) swimming with them. The fish kept so close that it was not difficult to catch both animals together. But once in the same container, the fish suffered and die within a few minutes although the jellyfish continued to live for more than 24 hours. Similar behaviour has been reported for the man-of-war fish when it is confined in a bucket with *Physalia*. The accompanying fish are, therefore, not totally immune to the sting. Perhaps the fish can tolerate the few strings which they may suffer in the wild, but cannot cope with the large battery of injections they suffer when they are closely confined with their host. A similar relationship exists between the clown fish and a tropical sea anemone. A pair of clown fish shelter among the tentacles of a particular anemone and they are protected from the stinging cells by a chemical found in the mucus on the body of the fish.

Most rhizostome jellyfish are found in the shallower warm waters of the world such as the Red Sea and parts of the Carribean and Gulf of Mexico, but a few species are found in temperate seas.

Class 3. The cubozoa — box jellyfish

These species are found in the Indo-Pacific, usually in shallow water, but they are occasionally seen in the open ocean (Map 9). They are easily recognised by their box-shape and the presence of one or a group of tentacles at each corner. They are notorious for their powerful sting; *Chironex fleckeri*, the sea wasp, has a translucent bell about 20 centimetres high, and this species has been responsible for at least 50 deaths in Australian waters. Even mild stings are painful and last for many hours, but the weals may persist for several months.

There are other species of box jellyfish which are slightly smaller, but their stings are still extremely powerful and the animals should be treated with respect.

SUBPHYLUM 2. ANTHOZOA — SEA ANEMONES AND CORALS

This group of coelenterates are polyps all their lives. The sea anemones are found all over the world on shores, pier piles and harbour walls, (Chapter 12; Colour Plate 62) but one species can be picked up in the plankton in warm tropical waters. Another anemone, *Anemonia sargassiensis*, as its name suggests, is commonly found among the *Sargassum* weed (gulfweed — Chapter 11). Corals, however, generally live only in clear, warm, shallow water (Map 10). Over 6,000 species go to build the world's coral reefs. Some have hard limestone, others are soft, but they are all basically like little anemones (the polyps) in an outer skeleton (Figure 4.6).

Figure 4.6: The structure of coral

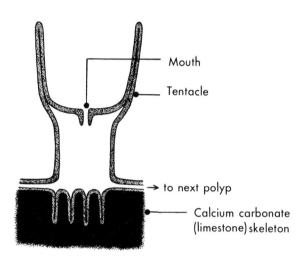

Mouth

Tentacle

→ to next polyp

Calcium carbonate (limestone) skeleton

The world's coral reefs and atolls were built over millions of years. Each atoll started as an island — usually volcanic — with a surrounding reef. Throughout the ages the land subsided and eroded away, and the sea rose, but the coral grew continuously, covering older dead animals to keep pace with the changing level of the sea. Eventually the island 'sank', leaving a lagoon enclosed by the reef — an atoll. The sea level has changed periodically, mostly due to the ice ages — melting ice raising and freezing ice lowering the level. Throughout history the coral has been repeatedly exposed and covered, until today the original island is often a mile below the reef, buried under many layers of coral which is now coral rock (Figure 4.7). Coral reefs like the Great Barrier Reef off the north-east coast of Australia have been built in a similar manner.

When coral is found on the shore it is brittle and white because the polyps are dead (although the holes in which they lived can be clearly seen). Living coral is brilliantly coloured, and reefs are magnificent underwater gardens. A visit to a local reef is always a worthwhile trip ashore, but it should be remembered that it takes millions of years to build, and collecting pieces will do irreparable damage. Taking coral is against the law in most coral reef areas of the world.

ATOLL

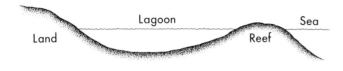

REEF

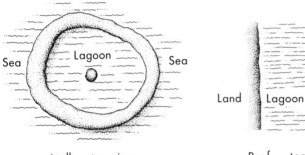

5. The Molluscs
Snails, Mussels, Squid and Octopus

Giant octopus and squid are the monsters of many seafaring stories, but they must remain in the imagination because although some giant squid may grow to lengths of 18 metres (60 feet) these are rarely seen alive. The body of the largest octopus grows up to 36 centimetres (15 inches) although the slender arms may extend up to five metres (15 feet). This species, *Octopus dolfleini*, lives on the Pacific coast of the USA.

Molluscs not only live in the sea but also on land, in rivers, and in lakes. The group includes snails and slugs; mussels and their relatives; the octopus, squid and cuttlefish. But despite the wide variety of body shapes, they all have basically the same structure: a soft body with a head and a muscular 'foot', and a skin called the **mantle** that covers the whole body. Many species have a shell which is made by the mantle and is either outside the animal to provide protection, as in mussels, or inside to give support as in squid. The octopus has lost its shell through evolution.

CLASSIFICATION

The classification of molluscs depends on the presence or absence, and type, of shell, and on the body shape.

Figure 5.1: Classification of molluscs

PHYLUM: CLASS:

Mollusca
— 1. Gastropoda — Snails with an outer shell (p. 62)
— Slugs with an inner shell (p. 66)
— 2. Bivalvia — Mussel types (p. 67)
— 3. Scaphopoda — Elephant tusk shell (p.67)
— 4. Cephalopoda — Squid, octopus etc. (p. 67)

CLASS 1. THE GASTROPODS

Snails and slugs have the stomach lying above the 'foot', hence the name gastropod — literally meaning 'stomach-foot'. The head has eyes and tentacles. Snails have a shell with an opening which can usually be closed with a 'door' once the animal is safely inside. Slugs have either a small internal shell, or none at all — merely minute spikes of calcium carbonate embedded in the skin.

1. SNAILS

Most snails have either a left-hand or right-hand coiled shell. Hold the shell with the opening towards you and the apex uppermost; if the opening is to the left of centre it is left-hand coiled, if to the right, right-hand coiled.

Hydrobia ulvae, (Fig 5.2)   *Hydrobia* is a black or brown snail between a half and three millimetres long which is often ·found at the water surface, suspended upside-down from a mucus raft. It feeds on particles that are trapped in the mucus, and fastens its eggs to the outside of the shell. These snails are common among the inshore plankton of estuaries at high water, and over mudflats.

1 mm

Figure 5.2: The snail Hydrobia ulvae *suspended from the surface film*

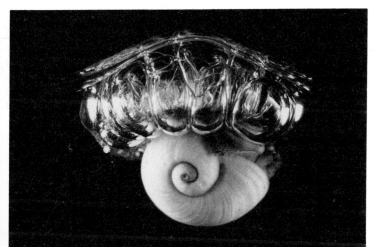

Plate 12: The floating bubble-raft snail, Janthina

The purple bubble-raft snails have violet or purple shells up to two centimetres high (Black and White Plate 12). They are suspended upside-down from a raft of bubbles which may be elongated, spherical or spiral and is constantly being built at a rate of about one bubble per minute. Different species of *Janthina* are found throughout the world, drifting in the warm and temperate surface waters (Map 11). They are occasionally found in the coastal waters of south-west Britain and Europe.

Janthina spp.

The snail feeds on other floating animals such as *Velella* and *Porpita*, and even *Physalia* (Chapter 4), but it will eat anything available like planktonic crustaceans, insects on the surface, and even young *Janthina*. When the snail comes into contact with its prey, such as a by-the-wind-sailor, it attaches itself and exudes a purple dye which is thought to be an anaesthetic protecting the snail from the animal's tentacles. The snail consumes the whole animal, eventually leaving behind only the transparent sail and a husk. Often, *Velella* washed up on deck or on the seashore may have several *Janthina* attached.

Many snails, including *Janthina*, are males during their early life, but later they develop into females. The mature male *Janthina* shed sperm into the water in packets, which enables them to survive the journey to the female that may be floating some distance away. The eggs are also shed in packets, and after fertilisation are attached beneath the raft of bubbles (Figure 5.3). There is no young stage, the eggs hatching directly into small versions of the adult.

Figure 5.3: The purple bubble-raft snail, Janthina sp.

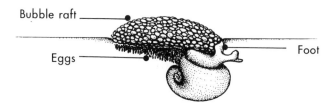

Bubble raft

Eggs

Foot

Other snails are even better adapted than *Janthina* to living in the oceans. The first of these groups consists of the swimming snails or **heteropods** which are flattened sideways so that the foot acts more like a fin as they swim upside-down or suspended from the surface film of the water. Some have a well developed but very delicate, thin, coiled shell. Heteropods are predatory, feeding on small fish and other planktonic animals.

Carinaria lamarcki

This snail has a thin transparent body up to 10 centimetres long. The 'fin' is uppermost and the tiny shell, with the gills sticking out, is carried beneath and acts as a keel (Black and White Plate 14).This species is found floating from the surface film in the warmer waters of the Atlantic and the Mediterranean, but other species of *Carinaria* are found around the world. Some migrate to the surface at night but sink to depths of around 100 metres during the day.

Plate 13: The swimming snail, Carinaria. The shell is at the bottom right, the mouth is on the upper right of the picture

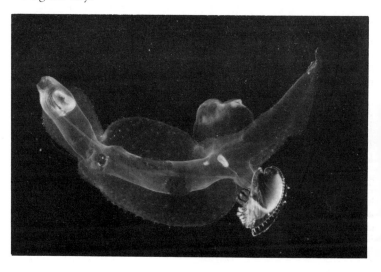

Atlanta spp.

This species lives in a well-developed thin spiral shell which has a keel running around the outside (Black and White Plate 14).

The second group of planktonic snails comprises the sea butterflies or **Pteropods**. The foot of these snails has evolved into two fins that look like butterfly wings. One group has a shell; the rest do not, and are called naked pteropods.

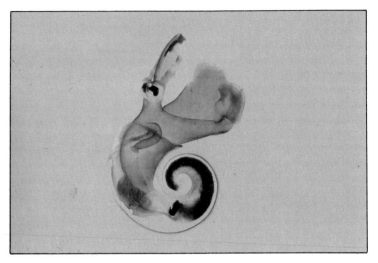

Plate 14: Another swimming snail, Atlanta. *The eyes are towards the top, and the whole animal can withdraw into the transparent, coiled shell*

Shelled pteropods

The shell is usually external and may be coiled, with a 'door', or of many different shapes without a door. Some have a false internal shell which can usually be seen as the only opaque part of the animal. They are found all over the world in the surface waters at dusk and dawn, but deeper during the day. They feed by trapping small planktonic organisms in mucus.

Cavolinia tridentata

This species is found world-wide. The shell is a distinctive triangular shape, with the three points drawn into fine projections (Colour Plate 23). The body is enclosed in the shell, with the opening near the apex of the triangle. They grow to about two centimetres long and swim using the two large fins, which flap like wings. A mucus coating on the fins traps plant plankton and other small particles of food. The mucus/food mixture is collected into strands by the synchronous beating of many fine hairs, and is pushed into the mouth between the fins.

Spiratella spp.

Spiratella has a small (between two and five millimetres high) left-handed coiled shell which is usually brown with lines or ridges running around the outside. These are possibly the most common of all the shelled pteropods.

Corolla spectabilis

The animal is almost entirely transparent except for the internal shell and a fine lattice-work of muscle fibres in the fins. It spins a mucus net to trap plankton for food and is found in the surface waters of the oceans.

Gleba cordata

Although the animal is usually between five and seven centimetres across the fins, it can spin a mucus membrane up to two metres in diameter. The snail dangles beneath the membrane, attached by the mouth at the end of a long snout: when the membrane is clogged with food it is sucked into the mouth. The animal is transparent with a thin, clear, glass-like shell and is found near the surface of the sea.

Naked pteropods   The members of this group have a muscular body without a shell. The head has tentacles, and the fins are relatively small and perform a rapid sculling action that enables the animals to swim quite fast. They live near the surface of the open ocean, often in large numbers during the day, feeding mainly on shelled pteropods caught by the tentacles, or by arms which have suckers, or on hooks that can protrude through the mouth.

Cliope limacina   This is perhaps the most common species living in the northern oceans. The elongated body is usually about two centimetres long, although Arctic species may grow to four centimetres (Figure 5.4). They have sticky tentacles as well as hooked sacs which can protrude through the mouth to catch the food. The head and tentacles are orange-red, the fins are pink and the internal organs appear brown and yellow.

Cliopsis spp.   These have small fins and are found in the surface waters throughout the oceans (Figure 5.4). They seize their prey with hooks that are pushed out from inside the mouth.

Figure 5.4: Two naked Pteropods, or sea butterflies

Cliopsis spp.

Clione limacina

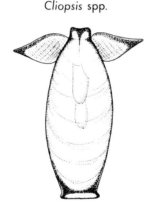

2. SLUGS

Three species of sea slugs are planktonic. Two can be found drifting on the surface; the third is usually found with *Sargassum* weed (Chapter 11).

Phylliroe bucephala   *Phylliroe* is a transparent flattened species up to two centimetres long with a pair of tentacles (Colour Plate 5). The internal organs can be seen as orange-brown structures. It is common in the surface waters of the North Atlantic but it has also been reported from the East Pacific and Indo-Pacific (Map 12).

Glaucus atlanticus   An oceanic species found world-wide in warm waters (Map 13). It is a distinctive black silvery animal growing up to five centimetres long with three bunches of flat finger-like projections on either side (Colour Plate 6). *Glaucus* glides along the surface film and devours floating coelenterates like *Velella* and *Porpita*. It is immune to the stinging cells of these animals.

The *Sargassum* slug lives among the Gulfweed found in the central North Atlantic and Gulf of Mexico (Chapter 11, Map 160). It grows to about six centimetres and is well camouflaged to resemble pieces of the weed although it may occasionally be found with other floating organisms, like the purple bubble-raft snail, **Janthina** sp.

Scyllaea pelagica — the *Sargassum* slug

CLASS 2. THE BIVALVES

These molluscs have a shell made of two halves, hinged together along one side (bivalve means two halves). In most species the body is completely enclosed in the shell. At sea, with one rare exception, they will only be seen in plankton samples as young stages (Chapter 2, p. 26 — the adults are either attached to the sea-bed or live in burrows in sand or mud. Occasionally they may be picked up on the anchor cable, but the can be found on most sea shores and are invariably among the animals that foul coolers, seawater intakes and condensers. Some species like *Teredo* and *Bankia* are commercially important because they burrow into wood; they are commonly called ship worms (Chapter 12).

CLASS 3. THE SCAPHOPODS — ELEPHANT TUSK SHELLS

These shells get their name from their elephant tusk shape. They live buried head-down in soft sea-beds in most warm waters of the world, and they are commonly found on the anchor cable. Their young stages may be picked up in plankton samples (Figure 2.10).

CLASS 4. THE CEPHALOPODS

The animals in this class include the largest and most complex of all the invertebrates. Some have an internal shell, others no shell. Many are able to swim very fast. They have no 'foot' but the head is well developed with prominent eyes; the mouth is shaped like a parrot's beak and is encircled by tentacles which usually have one or two rows of suckers.

The male usually has one tentacle (occasionally more) modified to carry a bag of sperm. In cuttlefish and squid the end of the right or left fourth tentacle has an area of small adhesive suckers; in the octopus the end of the third right tentacle is shaped like a spoon. When mating this tentacle is pushed into the female's mantle and the bag, full of sperm, is deposited. The octopus tentacle may detach and remain with the female (originally this was thought to be a worm parasite). The female releases her eggs, and the sperm leave the bag so that fertilisation may occur either inside or outside the mantle.

The smallest cephalopod is a species of cuttlefish which lives in rock pools and grows to 15 millimetres long (0.6 inches), but the giant squid at 18 metres (60 feet) is the largest invertebrate.

The *Nautilus* is the only animal in this class that has a true external shell. They live in tropical seas, mainly the Indo-Pacific

Nautilus spp.

(Map 14) and often hunt in groups at night. Occasionally they are found on the surface but they can swim to depths of 500 metres (1650 feet). The coiled shell, which is built in chambers, may be 20 centimetres across and is yellow to pink with brown or red stripes; the body is in the last and largest chamber (Colour Plate 34). The animal controls in buoyancy by regulating the gas content of the inner chambers. Often the shells are sold in curio shops, but collecting these beautiful animals should be stopped otherwise they will disappear into extinction.

The numerous tentacles around the mouth have no suckers and are used to catch small crustaceans and fish. They swim by expelling water through a tube or funnel underneath the body.

Sepia spp. — the cuttlefish

The cuttlefish are oval, up to 30 centimetres long, and have an internal shell — the cuttlefish 'bone', which is often seen floating on the surface or washed up on beaches. They swim slowly by undulating motions of the fin that extends around the body, or they can dart quickly by using a jet of water directed through a tube (Figure 5.5). Cuttlefish can change the colour of their skin to camouflage themselves against the background. They are found world-wide in shallow coastal waters.

Figure 5.5: The cuttlefish, Sepia spp.

SQUID

Squid occur all over the world in coastal and open waters. Most migrate to the surface from deeper water during the night. The body is torpedo-shaped, often with side fins, and the head has ten tentacles — eight short heavy arms and two long extending tentacles. Each has suckers and often hooks (Colour Plate 35). The internal shell is flat and lies along the animal's back. Many squid have a gland containing a brown or black fluid called 'ink' which is squirted out to confuse any hunting predators, and may contain chemicals that are objectionable to fish, anaesthetising their senses.

Many squid are responsible for extensive displays of bioluminescence (Chapter 3) and have luminescent organs (photophores). Some species have luminescent bacteria in little hollows in the skin.

Squid are able to cruise slowly using their fins and they can dart rapidly using a directed jet of water. They are the fastest of the invertebrates, feeding on fish and shrimps. Many species are likely to be seen; however the following are most distinctive.

Family: Onychoteuthids — the flying squids

These animals are small — up to 30 centimetres (1 foot) long — and can be seen gliding over the surface like flying fish. They produce a powerful jet of water to drive them up out of the water, and their prominent fins can keep them airborne for about 100 metres. One species, *Doscidiscus gigas*, can reach

16 mph (26 kph) through the air.

Squid may be found on the deck, particularly during rough seas. The specimen in Colour Plate 35, *Onychoteuthis banksi*, is found world-wide at the surface at night, but down to 500 metres (1650 feet) during the day. The bite of this species is toxic to man but the animal is sold for human consumption in certain parts of the world.

If these animals are seen at the surface they are usually dead. They live at depths greater than 300 metres (985 feet), swim slowly, and are the main food for the sperm whale (Chapter 10). They may reach a length of about 18 metres (60 feet).

Architeuthids — the giant squid

The members of the octopus group mostly live on the sea-bed in shallow water, although a few are oceanic. They have short rounded bodies, eight arms with suckers and, with the exception of one family, have no shell. They are active predators feeding mainly on crabs and molluscs which are cracked open using the strong beak. There are deep-water octopus which have fins like a web between the tentacles but little is known about these animals which live at depths of over 2,000 metres (6,500 feet).

Octopods

Several species may be found in rock pools and shallow waters of tropical reefs and sea shores. In the Indo-Pacific, one species to be wary of is the blue-ringed octopus, *Haplochlaena maculosa*; it is common under rocks and stones from low tide level to 20 metres deep (65 feet). Although only 10 centimetres long its bite is deadly and will kill a human within 15 minutes. It is easily recognised, with a brown body covered in small blue rings.

This is probably the only species that is likely to be seen from a vessel. The female has one pair of arms with flaps; each makes a half of the shell which eventually covers the rear of the animal and is used mainly to protect the eggs (although the adult may retreat inside). The shell resembles that of the *Nautilus* except that it is thinner, white, and has ridges (Figure 5.6). The smaller male does not have a shell but may co-habit with the female. This species is found all over the world in warm tropical waters.

Argonauta argo — the paper nautilus

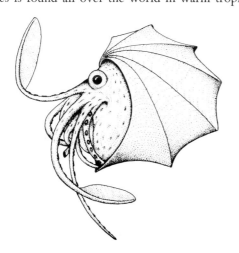

Figure 5.6: The paper nautilus, Argonauta sp.

69

6. The Crustacea
Barnacles, Shrimps, Crabs and Lobsters

Although crabs, shrimps and lobsters can be found on almost any sea shore and sea-bed, most of the crustaceans are small animals, forming the major part of the animal plankton (Chapter 2). Most crustaceans live in the sea but there are freshwater species and even some adapted to live on land though usually in damp places; for instance the wood-louse is usually found under tree barks or stones.

All crustaceans have an external skeleton, made from protein and hardened with calcium carbonate (limestone). The body is built in segments and divided into three parts: the head, thorax and abdomen. Many species have a cape or carapace covering the head and thorax (Figure 6.1). Every segment usually has a pair of limbs, each adapted for doing different jobs: food handling, crawling and swimming, transferring sperm, grasping the female during mating and for holding the fertilised eggs.

The adults have two compound eyes, that is each eye is made up of many smaller 'eyes' — the common lobster has up to 14,000 in each compound eye. Many crustaceans have gills attached to the tops of the legs and extending under the carapace.

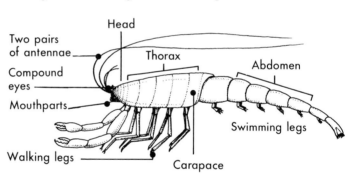

Figure 6.1: A generalised drawing of a crustacean: the head and thorax are together called the cephalothorax

CLASSIFICATION

There are over 30,500 species of crustaceans and they are all usually classified according to the body shape and the number and type of limbs.

Figure 6.2: Classification of crustaceans

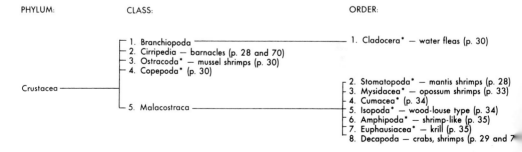

PHYLUM: CLASS: ORDER:

Crustacea
- 1. Branchiopoda ———————————————— 1. Cladocera* — water fleas (p. 30)
- 2. Cirripedia — barnacles (p. 28 and 70)
- 3. Ostracoda* — mussel shrimps (p. 30)
- 4. Copepoda* (p. 30)
- 5. Malacostraca
  - 2. Stomatopoda* — mantis shrimps (p. 28)
  - 3. Mysidacea* — opossum shrimps (p. 33)
  - 4. Cumacea* (p. 34)
  - 5. Isopoda* — wood-louse type (p. 34)
  - 6. Amphipoda* — shrimp-like (p. 35)
  - 7. Euphausiacea* — krill (p. 35)
  - 8. Decapoda — crabs, shrimps (p. 29 and 7

Note: Those groups marked with an asterisk (*) are discussed in Chapter two.

All marine crustaceans produce planktonic young (Chapter 2). Classes 1, 3, 4 and most of 5 include animals that are planktonic all their lives and were discussed in Chapter 2. This chapter is concerned with the larger animals: the barnacles, crabs, shrimps and lobsters.

CLASS 2: THE CIRRIPEDES — BARNACLES

Of all the crustaceans, barnacles are perhaps the most strange. They live firmly attached to a surface — rocks, shells, coral, floating debris, even whales, turtles and fish. Economically they are important, attaching to buoys, pilings, the legs of oil rigs, and ships; and they are the bane of an engineer's life, growing in coolers and condensers (Chapter 12). They may cause up to 30 per cent reduction in the speed of a ship, and the vessels must then go into dry-dock to be cleaned and painted — and everyone knows the amount of money involved in this operation!

There are three types of barnacles: those with a flexible stalk like a goose's neck and called goose barnacles; the second type have no stalk and are called sessile barnacles; the third type usually bear no resemblance to the other two — these are the parasitic barnacles.

Goose and sessile barnacles are basically little shrimps living inside limestone 'houses' (Figure 6.3). They have fine hairy legs (cirripede literally means 'hairy legs') which are pushed through an opening to catch small plankton for food (Colour Plates 36 and 37). The 'house' is made of calcium carbonate (limestone) plates. Goose barnacles usually have up to five plates around the main part of the body, but sessile barnacles have between four and eight plates built in a low heavy circular wall around the whole animal. The individual plates may be fused so they are indistinct, or they may be clearly marked.

Figure 6.3: The two types of barnacle in their 'houses'

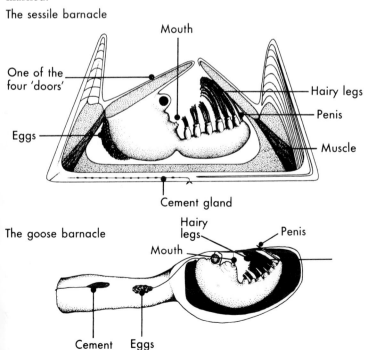

The sessile barnacle

- Mouth
- One of the four 'doors'
- Hairy legs
- Penis
- Eggs
- Muscle
- Cement gland

The goose barnacle

- Hairy legs
- Penis
- Mouth
- Cement gland
- Eggs

The parasitic barnacles have various body structures, but generally males and females are separate individuals whereas most other barnacles are **hermaphrodite** — that is each individual has both male and female sex organs which produce sperm and eggs. The sperm of one animal is deposited through the opening of a neighbour via a long penis. This is one reason why barnacles tend to live close together. The eggs are kept inside the house before they hatch into **nauplius** stages which swim off in the plankton (Chapter 2, Figure 2.11). Each barnacle can produce up to 13,000 offspring.

After six moults the nauplius changes into a **cyprid** (Figure 2.11) which does not feed but searches for a suitable place to settle, where it will attach, moult and begin its adult life. The cyprid stage barnacle is attracted to areas where barnacles have already settled, or where they have settled in the past. The barnacles leave behind a chemical, stuck indelibly on the rock, which is resistant to weathering and can be detected by the cyprid. The presence of the chemical is an indication that barnacles have survived there, hence it is likely to be a 'good' place to settle. It is also important for the cyprid to detect the presence of other barnacles because they must be close to each other to be able to mate. If a substrate is found to be unfavourable the cyprid will swim off and try elsewhere until there comes a time when the barnacle must settle on whatever substrate is available — whether suitable or not.

Gradually the cyprid sticks itself to the substrate using a strong 'cement' produced by a gland in the antennae. The young barnacle moves onto its back and builds the limestone plates into its shell-like house. (This is not the same as the outer skeleton which lines the inside of the house covering the body and legs, and is moulted as the animal grows.) Once settled, the barnacle may live between two and six years depending on the environmental conditions. The animal constantly extends its house by adding more calcium carbonate, and cement is added to the base throughout the barnacle's life so they can usually repair themselves if they are partially detached, although removal from the substrate usually means death at the hands of scavenging crabs or gulls. However, the main cause of death is smothering by other barnacles and other attached animals.

Barnacles can be collected from the engine room whenever a cooler, condenser or any other seawater container is being cleaned. There are likely to be many species, depending on where the ship has been since the cooler was last cleaned. Some barnacles are small, others large; some have little horns on the outside, others are smooth; there are plain or striped and multi-coloured varieties. Often the animals are stained with rust and sometimes they are in poor condition because of the difficult environment in which they live. These specimens may disintegrate as you try to take them off, but occasionally the animals can be removed whole. If they are put in fresh seawater they soon settle down and can be seen pushing their legs — like a tiny grasping hand — into the water, collecting plankton for food (Colour Plates 36 and 37).

1. Sessile Barnacles

Some sessile barnacles are only a few millimetres in size and burrow into shells and coral, but most are about one centimetre in diameter and live on rocks, shells and other surfaces. The largest can be found on the west coast of South America. This species, *Balanus psittacus*, grows to 23 centimetres (9 inches) high and 8 centimetres (3 inches) in diameter.

These barnacles are well adapted to living on sea shores, jetties and pier piles. Their large area of attachment and low circular wall enable the animals to resist heavy breaking waves and strong currents. There are three species that grow on particular animals. They are not parasites, as a parasite is an animal or plant which harms the organism on which it is living. These barnacles do not harm their host, they are simply using it as a place to live.

Chelonebia spp. live on the backs of turtles and swimming crabs.

Coronula spp. live on whales, particularly the California grey whale (Chapter 10).

Xenobalanus spp. look more like a stalked barnacle and live on the fins of whales and dolphins.

2. Stalked or goose barnacles

Goose barnacles are not as common in the engine room although they may be collected inside large seawater condensers (Colour Plate 38). They are mainly found attached to drifting debris, gulfweed, turtles and whales and ships which generally pick up these barnacles away from coastal water. Some species are attached to rocks on the sea shore and in shallow waters of the Mediterranean and the North-East Atlantic and Pacific coasts. (The stalks are considered a delicacy in Mediterranean countries.) A few species of goose barnacle are usually found in *Sargassum* (Chapter 11).

Lepas anatifera

This species is found all over the world attached to floating objects. It has five plates around the body, which is at the end of a flexible stalk up to 25 centimetres (10 inches) long.

Lepas anserifera

L. anserifera is similar but has a smaller stalk and the plates have distinctive furrows (Colour Plate 37). They are found in the warmer waters of the world.

Lepas fasicularis

This goose barnacle can build its own float from spongy bubbles, but the young must first attach to a small floating object. As the animal grows, and other barnacles become attached, the float is added to keep the group buoyant. The body grows to about four centimetres long and has five translucent plates.

Conchoderma spp.

The *Conchoderma* spp. have less than five plates, and these are small and separate from each other. One species has only two plates. They are commonly found growing on turtles, whales, floating debris and ships' hulls (Colour Plate 38).

3. Parasitic barnacles

Parasitic barnacles attach to or burrow into another animal, and in some way do it harm. These barnacles have no 'house', limbs or gills because they get protection, food and oxygen from their host. Figure 6.4 shows the life cycle of one species that attacks crabs. A young female barnacle burrows into the crab through the joint between two segments and grows inside until, when sexually mature, she pushes out a growth from the crab's abdomen. This is a chamber for holding the eggs (Colour Plate 39). A young male barnacle attaches itself to the chamber and sheds cells into the female, which are stored and eventually develop into sperm. The eggs hatch into either male or female nauplii which eventually moult into cyprids, and the cycle continues.

Figure 6.4: The life-cycle of a parasitic barnacle (see also Colour Plate 39)

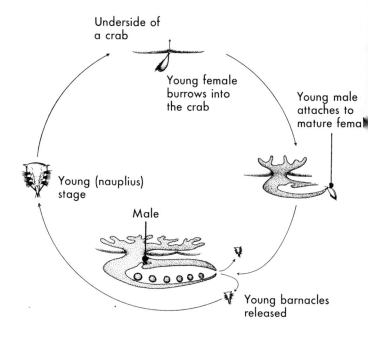

Underside of a crab

Young female burrows into the crab

Young male attaches to mature female

Young (nauplius) stage

Male

Young barnacles released

The effects on the crab are drastic: moulting is stopped, and it is castrated and so cannot reproduce. Other parasitic barnacles attack starfish, sea urchins and corals — different species being parasites on different animals.

CLASS 5. THE MALACOSTRACANS

Most of these animals are small and planktonic, but this class also includes the largest crustaceans.

Order 8. The Decapods — crabs, shrimps and lobsters

Decapod means ten legs, and all the animals in this group have five pairs of limbs. Usually the first pair are armed with pincer-like claws and the remaining pairs are for crawling and swimming. There are over 8,500 species ranging from the smallest crab, a few millimetres across, to the Japanese spider crab, *Macrocheira kaempferi*, the largest living crustacean, with a body 45 centimetres (nearly 18 inches) long and the pincer limbs reaching up to four metres. (Some lobsters may have bodies up to 60 centimetres, about 2 feet, long.) A few decapods live in fresh water, some are amphibious and others live on land, but most are marine — usually living on the shore or sea-bed.

These animals have no distinct division between the head and the thorax; both are hidden beneath the carapace, which also protects the gills. Lobsters and shrimps have a well-developed abdomen with a tail fan, but the crab's abdomen is almost unnoticeable folded up underneath, between the last pair of legs (Colour Plate 39).

Decapods have an infinite variety of colours, and many shrimps and crabs can change their colour to match their surroundings. For instance, the *Sargassum* shrimps and crabs are all olive-green to yellow in colour, with a mottled appearance (Chapter 11) to blend in with the weed. The decorator crabs, members of the spider crab family, attach weeds, sponges and sea firs to their backs and legs for camouflage.

Most decapods crawl on the sea-bed or shore. They are the 'dustbinmen', scavenging on dead and dying animals; some are carnivorous and attack other smaller animals; a few filter plankton from the water. Crabs are sometimes found in the engine room, usually in the main seawater intake filter.

Many decapods have a courtship ritual to attract sexually mature individuals. In some species the female releases a scent to attract the male. Usually males have one of the abdominal legs adapted for transferring sperm, and in some species this resembles a syringe which pumps the sperm into the female. Other species produce a bag to store the sperm and the modified leg is in the form of a clasper, used to plant the bag onto the female's receptacle. Most species carry the fertilised eggs attached to the female's abdomen until they hatch as **zoea**. Eventually these moult into **megalopas** that resemble the adults (Chapter 2, Figure 2.12). There are variations in this pattern of growth depending on the species.

After spending a certain amount of time swimming in the plankton, the young decapod completes its last moult, emerges as a young adult, and settles into the adult life. Even land crabs migrate to the sea to lay their eggs; one species walked eight kilometres to the shore to lay eggs.

Crabs and shrimps can be found on the sea shore: shrimps are more likely to be in rock pools from about mid-tide to the low water mark; crabs may be found scuttling over rocks or hiding under stones or among seaweed. Lobsters are usually in deeper water, and seafarers are only likely to see them in local restaurants or fish markets!

Ghost crabs are found on tropical and sub-tropical sandy beaches, particularly in the Caribbean and Indo-Pacific. They are about five centimetres across and live in deep burrows above the high-water line, often among the sand dunes. At night they emerge to feed on any debris they can find on the beach. They run sideways, occasionally turning the body through 180° to alternate the leading and trailing legs. Some species reach speeds of more than one and a half metres per second (about three and a half miles an hour). They return to the water if they are disturbed, and to moisten their gills.

Ocypode spp. — the ghost crabs

Like the ghost crabs, fiddlers live in burrows on tropical shores, but usually in muddy areas above and below the high-water line. They also emerge to feed on debris left behind by the ebbing tide. At high tide they hide in their burrows and plug the entrance with a ball of sand. The male has one large claw which is used in a courtship display to entice the female to his burrow.

Uca spp. — the fiddler crabs

Family: Grapsids — the climbing crabs	These species are also found in tropical and sub-tropical regions. They grow to about five centimetres across and climb rocks and mangrove trees, clinging on with spines at the ends of their legs (Colour Plate 40). They are usually camouflaged with a dark green colour and are therefore difficult to see against wet rock.
Family: Gecarcinids — the land crabs	Land crabs grow to about fifteen centimetres across and live in tropical areas — particularly in the West Indies. They can be found well beyond the high-water mark and use either fresh or sea water to moisten their gills (Colour Plate 41). They do not feed on the beach but the female always returns to the sea to lay her eggs.
Family: Portunids — the swimming crabs	These are the most powerful and agile swimmers of all the crustaceans. The last pair of legs are flattened like paddles (Colour Plate 42), and each performs a 'figure-of-eight' movement in opposite directions. This counter-beating acts like a propeller and pushes the animal through the water, the other legs acting as stabilisers. Some species can swim fast sideways, backwards and forwards, and can even catch small fish.

Swarms of small pink swimming crabs between five and ten centimetres across have been seen in the Gulf of Oman and in the tropical Atlantic and Pacific off the coast of South America (Map 15). They probably belong to the *Charydris* group of species.

7. The Urochordates
Sea Squirts and Salps

Sea squirts are strange little animals living inside a skin which is like a closed bag with two small tube-like openings, one to let water in, the other to squirt it out (Figure 7.1). Many species are found firmly attached to rocks on the sea shore, or to pier piles, harbour walls and ships' bottoms; others swim freely in most waters of the world. Some species have been dredged up from as deep as 2,000 metres (6,500 feet).

They are very common although not easily recognisable as living animals. The skin, or tunic, may be coloured but it is usually transparent — particularly in the swimming species. This makes them difficult to see in the water, but when taken out they appear like flaccid lumps of jelly and may be mistaken for small sausage-shaped jellyfish.

They are simple animals; water is sucked in through the inlet, passes through a basket network, and out of the outlet. Small animals and plants are trapped in mucus in the basket and are used for food. Oxygen is also taken from the water. Attached sea squirts have the openings pointing up, but the swimming species have the inlet at the front and the outlet at the rear; the resulting current produces a 'jet' action pushing the animal through the water.

Figure 7.1: Generalised diagrams of the two types of sea squirts

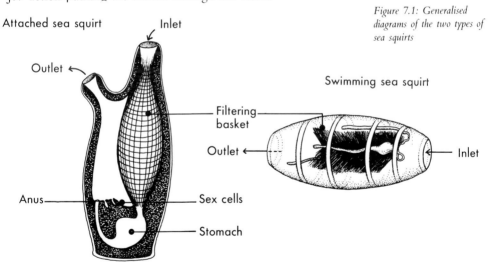

Attached sea squirt   Inlet

Outlet ←

Outlet ←

Filtering basket

Swimming sea squirt

Outlet ←   → Inlet

Anus

Sex cells

Stomach

Sea squirts have male and female sex cells in the same animal — they are hermaphrodite. The eggs are fertilised inside the body, or after they have been shed into the sea: in either case a tadpole-like planktonic young is produced (Chapter 2; Figure 2.21).

CLASSIFICATION

Sea squirts may appear like insignificant lumps of jelly, but they are man's distant relatives because they have a very simple 'backbone'.

This may be present only in the young stage, but nevertheless it is significant. Having this structure puts these animals on the dividing line between those animals without a backbone — the invertebrates, and those with — the vertebrates.

Figure 7.2: Classification of sea squirts

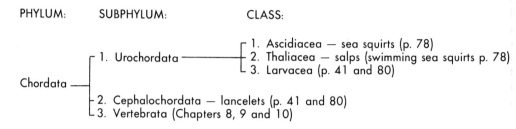

The chordates are a large group of animals. The urochordates and cephalochordates are discussed in this chapter; the vertebrate groups — fish, reptiles and mammals — are discussed in the following three chapters.

SUBPHYLUM 1. UROCHORDATES

CLASS 1. THE ASCIDIACEANS — THE SEA SQUIRTS

These are true sea squirts, living attached to submerged objects or even other animals such as crabs. Occasionally they may be found on floating debris or on ships' bottoms, but they are most frequently seen attached to rocks, harbour walls and pier piles at about the low-water mark.

After fertilisation the planktonic tadpole young is produced which possesses the significant rod or simple 'backbone'. Eventually the 'tadpole' settles and attaches to an object; it absorbs the backbone and tail (similar to the frog tadpole) and changes into an adult.

There are over a thousand different species. Some live as single, solitary individuals (Colour Plate 43) others grow buds that develop into many new individuals which eventually create a colony (Colour Plate 44).

Ciona intestinalis is perhaps the most cosmopolitan species. The body is a soft rubbery tube growing up to 15 centimetres high and often with yellow or green hues; the openings are occasionally fringed with yellow.

CLASS 2. THE THALIACEANS — THE SALPS OR SWIMMING SEA SQUIRTS

Salps are swimming sea squirts with the openings at opposite ends of the body so the water flowing through provides 'jet propulsion'. Solitary species look like small, transparent, pulsating barrels, particularly as many have rings of muscles running around the body (Black and White Plate 9). Other species live in colonies, which can sometimes grow very large (Colour Plate 45).

The majority of salps are found in the sub-tropical and tropical areas of the world, although some species live in cooler waters. Certain salps are responsible for particularly dramatic displays of bioluminescence (Chapter 3).

In many cases the life-cycle is complicated involving sexual reproduction and non-sexual budding where many individuals grow from one (Figure 2.21).

Pyrosoma is a hollow cylindrical colony of individuals sharing one large skin. Each individual's inlet points outward from the cylinder, but the outlet points to the middle (Figure 7.3). The cylinder has the back end open; water is sucked in by each individual salp and passes through into the cylinder and out of the back, thus pushing the colony along. Although the individuals may be fairly small the colony may grow to over two metres long and up to a metre in diameter (Colour Plate 45).

Pyrosoma spp.

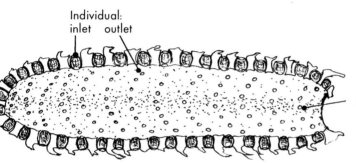

Individual:
inlet   outlet

Colony outlet

Figure 7.3: Diagram of a cylindrical Pyrosoma *colony cut in half*

Each individual is hermaphrodite and the eggs are fertilised internally. There is no young stage — the fertilised eggs stay inside the parent colony until they can swim free as young adults. These then produce buds which will develop into the new individuals to build a new colony.

Pyrosoma has luminescent organs inside each individual and these colonies are known to be the cause of some extremely dramatic displays of bioluminescence. It is possible to write your name in the luminescent slime on the body of one of these large colonies. The animals will also react to a searchlight, giving off a very bright glow. It has been reported that observers have been able to 'write' on the water by using a beam of light. This has been attributed to the presence of many colonies of *Pyrosoma*.

Species of *Pyrosoma* are mainly found in the sub-tropical and tropical seas, but the larger specimens have been reported mainly from the southern oceans (Map 16).

The doliolids are transparent, barrel-shaped animals growing to about a centimetre. The tunic is thin and there are eight or nine distinct muscle bands completely encircling the body. The complicated life cycle, starting with the tadpole young, is described in Chapter 2, Figure 2.21. These animals are cosmopolitan, occuring in the plankton of both tropical and temperate waters.

Doliolids

These creatures are similar in shape to doliolids, but they have a

Salps

thick tunic and the muscle bands do not form complete rings around the body. The animals vary from a few millimetres to several centimetres long but they may be joined together in chains up to a metre long. The fertilised egg develops inside the parent until the young hatches and grows the fleshy tail; then it is expelled through the outlet. The tail divides into segments, each developing into an individual so that a chain of small salps is formed. Three types of salps may be found in all the warmer waters:

Cyclosalpa spp. are up to eight centimetres long. The young are joined in a circle connected to a fleshy tail beneath the animal (Black and White Plate 9).

Pegea confeoderata may grow to 12 centimetres long. It has very short muscle bands barely reaching the sides of the body.

Thetys vagina is perhaps the largest salp, growing to 20 centimetres long. It is broader at the front than at the back where there are a pair of limb-like outgrowths. The body surface is covered by small pointed lumps.

CLASS 3. THE LARVACEANS

These are a strange group of animals that look like young sea squirts in the tadpole stage. They have, in fact, developed from sea squirt young by taking on the adult characters while staying in the plankton. They are discussed in Chapter 2 (Figure 2.22).

SUBPHYLUM 2. CEPHALOCHORDATES — THE LANCELETS

There are only 25 species in this group. They are small, translucent fish-like animals five or six centimetres long, with a narrow fin running along the back, round the tail and half-way along the underside (Figure 2.23). The mouth is surrounded by small tentacles; there are V-shaped muscle bands throughout the body, and the animal swims by quick body flexures.

There are males and females, but the eggs and sperm are shed so that fertilisation occurs in the water. The young are small, transparent and planktonic, migrating to the surface at night but down to the sea-bed during the day. The adults can be found in coarse sand and shell gravel, feeding on small animals filtered from the water. In certain parts of the world the adults tend to congregate in large numbers and they are fished commercially.

8. The Fish

The sea teems with an infinite variety of fish with different shapes, sizes and colours — the notorious sharks, the simple flying fish, the flashing swordfish and countless others. They all have special senses with which to obtain information about their surroundings: the lateral line detects water pressure, telling the fish of a nearby animal; they have a well-developed sense of smell; some fish can generate their own electricity.

This chapter serves as a guide to the common families of fish, but for correct identification of the species further reading is necessary. However, those families and species that are commonly spotted at sea are discussed in some detail.

CLASSIFICATION

The length, fin size and the presence or absence of fins and other structures are all important points for correct identification. Figure 8.1 shows the general classification, and the general outlines in Figures 8.4 and 8.7 are intended as a quick guide to the major families.

SUB-PHYLUM: CLASS:

Figure 8.1: Classification of fishes

Vertebrata ——————
- 1. Cyclostomata — hagfish and lampreys (p. 81)
- 2. Chondrichthyes — sharks, skates and rays (p. 82)
- 3. Osteichthyes — bony fish (p. 87)

CLASS 1. CYCLOSTOMES — HAGFISH AND LAMPREYS

These are primitive eel-like fish with a round or slit-shaped mouth, and horny teeth (Figure 8.2). They have no bone, and the body is supported with pieces of cartilage. The skin is slimy, especially in the hagfish which can fill a two-gallon bucket with slime and water in a few seconds.

Figure 8.2: A hagfish and a lamprey

Hagfish

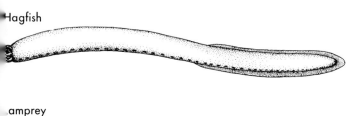

Lamprey

Lampreys ascend rivers to spawn, and the young spend their early lives feeding on rotting material in the river. The adults are parasites on fish and mammals, attaching themselves with the mouth and sucking body fluids from their host, then leaving behind a round scar. Hagfish are entirely marine and feed on any dead or dying animal that falls to the sea-bed. Both these animals are found in all temperate waters.

CLASS 2. CHONDRICHTHYES — SKATES, RAYS AND SHARKS

The fish in this group are easily recognised by their distinctive shape. The mouth lies beneath the head, and there are several gill slits on either side (Figure 8.3). The skin is covered with small tooth-like structures, called denticles, embedded at the base and ending in sharp points. These are directed backwards and produce the characteristic feel of sharkskin. The skeleton is made of cartilage, although some of the vertebrae may have calcium carbonate on the surface. These fish have an extremely well developed sense of smell.

Figure 8.3: Generalised diagrams of sharks and skates (or rays)

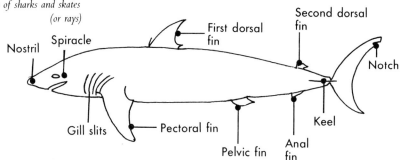

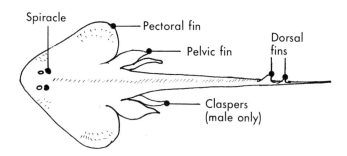

Fertilisation is internal, and mature males have special clasper organs on the inside surfaces of the pelvic fins. Each clasper has a canal which carries sperm into the female opening. The young are either born alive or each egg is laid in a brown capsule called a mermaid's purse, which may occasionally be found washed up on the shore.

Sharks are notorious but other members of the group also need a word of caution. Stingrays are often found on sandy bottoms, especially in warmer waters. The tail bears a venomous spine which, although rarely fatal, is intensely painful and may have a paralysing effect. The wound should be washed in saltwater and then plunged

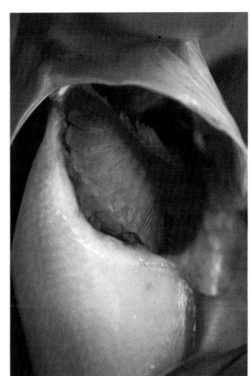

26

27

28

29

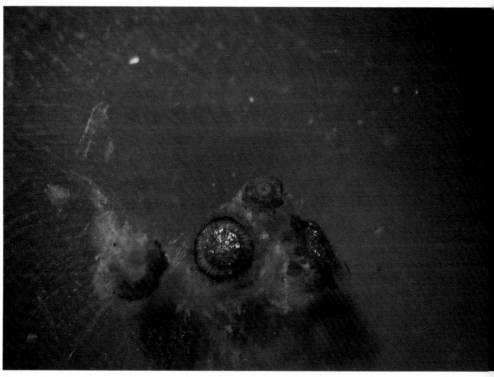

30

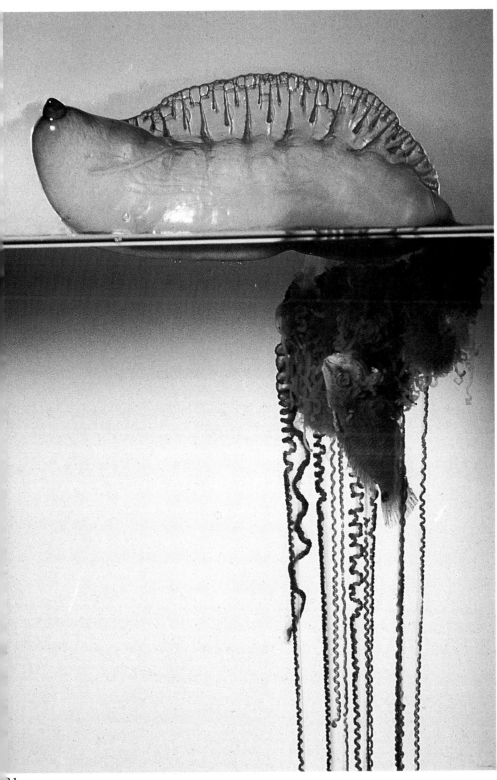

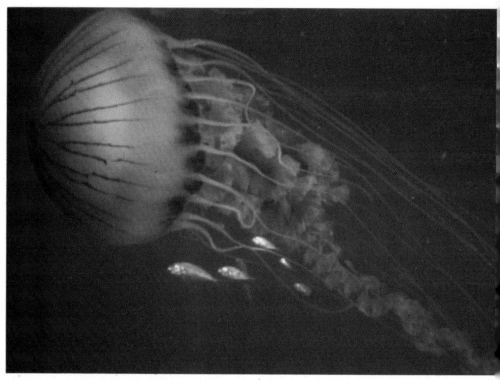

32

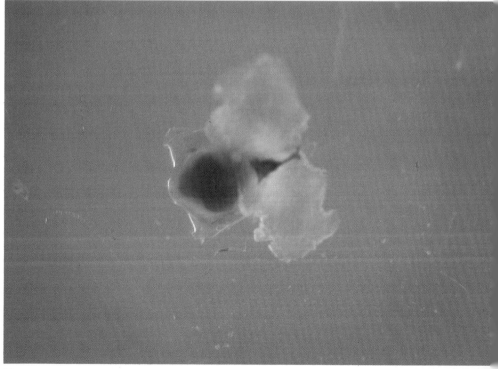

33

34

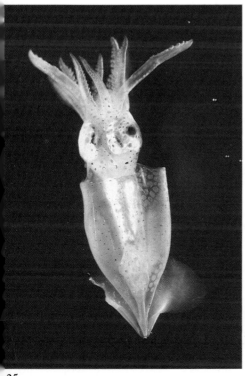

35

36

37

38

39

40

41

43

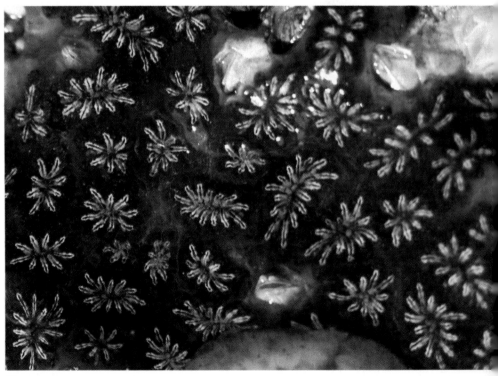

44

into water as hot as possible. When the pain decreases any piece of the barb still in the wound should be removed.

Electric rays should also be treated with respect. Some species can develop a voltage as high as 220v although it is probably much lower underwater. The animals use the shock to stun fish for food. A fisherman may receive a shock from his line after hooking an electric ray.

Figure 8.4 shows eight families of sharks and rays that may be commonly seen.

Figure 8.4: A pictorial key to eight families of sharks and rays

Family 1: Rhincodontidae
— whale shark (p. 83)
Size: 15 metres

Family 2: Cetorhinidae
— basking sharks (p. 84)
Size: 12 metres

Family 3: Alopidae
— threshers (p. 84)
Size: 5 metres

Family 4: Isuridae
— mackerel sharks (p. 84)
Size: up to 6 metres

Family 5: Carcharinidae
— requiem sharks (p. 85)
Size: Mostly around 2 metres

Family 6: Sphyrnidae
— hammerheads (p. 85)
Size: 4 metres

Family 7: Mobulidae
— manta rays (p. 86)
Size: wingspan 6 metres

Family 8: Myliobatidae
— eagle rays (p. 87)
Size: wingspan 2.5 metres

The one species in this family, *Rhincodon typus*, is the world's largest fish, growing to 15 metres (50 feet) long (Colour Plate 7). But, despite its size, it is harmless to man and feeds on plankton filtered from a constant stream of water which flows into the mouth, through the gills, and out of the gill slits. The head is blunt when viewed from above, and there are many tiny teeth

Family 1: Rhincodontid — the whale shark

— about 15,000 — which are not used for biting but for holding in the contents of the mouth. Several ridges run down the back, which is dark and mottled. The underside is a light colour. The eggs are laid in cases, and one found off Texas measured 70 by 50 centimetres (27 inches by 20 inches).

Whale sharks are lethargic animals, which might explain their occasional collisions with ships. The animals live in the tropical seas in all three major oceans, but they are not common.

Family 2: Cetorhinids — the basking sharks

There may be several very similar species of basking shark. These large fish grow to 12 metres (40 feet) long, and have five gill slits extending most of the way round the body. The head is blunt, with small eyes and teeth; there are two dorsal fins and a large tail fin; the back is grey fading to a lighter colour underneath (Colour Plate 8).

Like the whale sharks, basking sharks are harmless plankton-feeding giants, taking food out of the water as it flows through the mouth and gills. They live mainly in open water, usually in shoals of 50 or 60, but migrate towards the shore in summer. During the winter the shark retreats to deeper water and hibernates.

Not much is known about their breeding but it is thought that they give birth after a pregnancy of up to three and a half years. Basking sharks occur in the north and south Pacific and Atlantic and in the Indian oceans (Map 17). One species, *Cetorhinus maximus*, lives in the Atlantic north from Madeira. It is thought that their partly decayed bodies washed up on beaches provided the basis for giant sea-serpent stories.

Family 3: Alopids — thresher sharks

These sharks have a long upper lobe to the tail fin that may be over half the total length of the body (Colour Plate 9). The total length of the animal is about 5 metres (16 feet). They are slender and dark blue, brown or black, shading to white underneath.

Threshers are usually seen singly, rarely in pairs, and usually offshore. They are surface sharks, feeding on fish such as herring and mackerel, which are herded into tight shoals using the long tail. The sharks have also been seen to use the tail to stun birds sitting on the surface. They give birth to two or four young, usually in shallower water.

The thresher, *Alopias vulpinus*, occurs in tropical and temperate waters. In the Atlantic it is found from Ireland to the Cape of Good Hope, including the Mediterranean, and from Nova Scotia to Argentina. In the Pacific it is known from Oregon to Chile and off Hawaii, Japan, China and Australia (Map 18).

The big-eyed thresher, *Alopias superciliosus*, occurs in deeper tropical waters around California in the Pacific and around Florida and Cuba north to Madeira in the Atlantic (Map 19). This species has much larger eyes, a slightly smaller tail, and the dorsal fin is set further back.

Family 4: Isurids — the mackerel sharks

This family of large oceanic sharks includes the mako, *Isurus oxyrinchus*; the great white, *Carcharodon carcharias*; and the porbeagle, *Lamna nasus*.

The mako is a bluish-grey shark which grows to 4 metres (13 feet) long, and has a pointed head and slender body. It is a fast-swimming surface shark, feeding on shoals of fish and often leaping out of the water in the chase. It occurs in the tropical and warmer waters of the Atlantic and Indian Oceans (Map 20). The Pacific species is the short-fin mako, *Isurus glaucus*.

The notorious great white shark grows to six metres (20 feet) and has a deep body, slate-blue to grey-brown in colour shading to white underneath (Colour Plate 10). It roams all tropical and temperate seas, although nowhere is it common (Map 21). It is a voracious eater having large triangular-shaped teeth with serrated edges, and feeds mainly on fish, turtles, seals and refuse from ships. Its teeth have been taken from the hulls of boats and it has attacked bathers.

The porbeagle is smaller than its relatives — up to 3 metres (10 feet) — but it is more thick-set. It is a dark grey-blue colour on its back, and white underneath, and occurs in the colder temperate waters of the Atlantic from Nova Scotia to South Carolina and from Iceland to the Mediterranean (Map 22).

Family 5: Carcharinids — the requiem sharks

This is the largest family of sharks living mostly around inshore tropical and sub-tropical waters although two species — the blue shark, *Prionace glauca*, unmistakable because of its brilliant blue colouring, and the tope, *Galeorhinus galeus*, a one and a half metre long dusky coloured shark — are common in European waters.

Most sharks in this family are around two metres (six feet) long, the exception being the tiger shark, *Galeocerdo cuvier*, which grows to 6 metres (20 feet). The young tiger shark has dark stripes or spots and is therefore easy to recognise. However, these fade as the shark grows, and the adults are a uniform brownish-grey colour. They are among the most voracious of sharks; they are cannabilistic, and will eat anything. Their stomachs have yielded cans, bottles, clothes, shoes, half a crocodile and even undetonated explosives (Map 23).

Perhaps the commonest shark of open waters is *Carcharinus longimanus*, the white-tip shark, so-called because of the pure white or sometimes greyish colour on the tips of the fins. The rest of its chunky body is light grey or brown, but pale underneath (Colour Plate 11). Most white tips are between two and three metres long and are rarely seen near the shore, spending their lives in the tropical and sub-tropical west Atlantic and east Pacific oceans (Map 24).

Family 6: Sphyrnids — the hammerhead sharks

These sharks are instantly recognisable because of the flat lobes on either side of the head, which give the characteristic hammer shape (Colour Plate 12). The nostrils are on the front corners and the eyes on the sides of the 'hammer'. There are several species, distinguished mainly by the shape of the head (Figure 8.5), their size and colour.

All hammerheads are found in sub-tropical and tropical inshore waters. They gather together in the summer months both inshore and in deeper water for a massive northward migration along the Atlantic and Pacific coasts of America, and the west and north-western coast of Africa. Occasionally, individuals may wander as far north as the European coast.

Figure 8.5: The head outline shapes of four hammerhead species

Sphyrna zygaena
— common hammerhead

Sphyrna tudes
— great hammerhead

Sphyrna lewini
— scalloped hammerhead

Sphyrna tiburo
— bonnethead

Sphyrna zygaena — the common hammerhead

This species has a smoothly rounded front but is deeply indented around the nostrils. It is dark olive to brownish grey, white underneath, and grows to 4 metres (13 feet). The common hammerhead is widespread in the tropical Atlantic and Mediterranean and occurs occasionally as far north as Europe (Map 25).

Sphyrna tudes — the great hammerhead

The great hammerhead has the front of the head indented in the centre, and the body is heavier and longer — growing to four and a half metres (14 feet). The species occasionally occurs in the Mediterranean (Map 26).

Sphyrna lewini — the scalloped hammerhead

This shark has large indentations at the centre and either side of the front of the head. It is light grey above, white below, and grows to three metres (ten feet). It occasionally occurs in the Mediterranean (Map 27).

Sphyrna tiburo — the bonnethead

The bonnethead is perhaps the commonest of shallow water sharks. The head is shovel-shaped, and the body, which is greyish brown or grey, grows to two metres (six feet). This shark is found in the warmer waters of the Atlantic and Pacific, occurring from North Carolina to Brazil and from California to Equador (Map 28).

Family 7: Mobulids — the manta rays and devil fish

These giant acrobats are unmistakable. They are often seen leaping out of the water, somersaulting, then landing with a loud smack on the surface. Mantas are harmless giants growing to a wing-span of 6 metres (20 feet) and spending their lives gracefully 'flying' through the water feeding on plankton and shoals of small fish. Sometimes they just bask lazily on the surface with their wing-tips turned upwards. The front fins project forward like horns and are used to funnel food into the mouth. The eyes are on the sides of the head, the teeth are very small, and there is a small fin towards the rear. Mantas do not lay eggs but give birth. They are found throughout the tropical and sub-tropical seas.

This impressive fish looks like a broad diamond-shaped disc from above. The mouth is on the underside, which is completely white; the topside is dark brown or black. The devil fish grows to a wing-span of up to 5 metres (16 feet) although a smaller relative, found in the tropical Atlantic, may be only 1.5 metres (5 feet) wide.

Mobula mobular — the devil fish

There may be two closely related species, the Atlantic manta, *Manta birostris* (Map 29), and the Pacific manta, *Manta hamiltoni* (Map 30), but there are few differences (Colour Plate 13). The mouth is at the front of the head; the back is dark blue to black but may occasionally have white markings or blotches on the shoulder; the underside is white. They grow to a wing-span of about 6 metres (20 feet).

Manta spp. — the manta rays

The eagle rays resemble the mantas but are smaller, growing to about two and a half metres (eight feet) across, and do not have the horn-like front fins. There are several species. All are diamond-shaped, with a long whip-like tail with one or two poisonous spines (Colour Plate 14). Like mantas, they are often seen leaping above the surface but their main food consists of bottom-living animals such as shellfish and crabs; some species are a pest on oyster farms. They have between three and seven young in each litter and live mainly in tropical seas but occasionally in European waters.

Family 8: Myliobatids — the eagle rays

These fish are brown, grey, greenish, or yellowish above and white below (Map 31).

Myliobatis spp. — eagle rays or bat rays

These rays have regularly arranged spots of white, yellow or green on the back. The tail is black and the underside white. They occur in the Atlantic from Angola to Cape Verde and from Brazil to Chesapeke Bay including the Caribbean Sea (Map 32).

Aetobatus spp. — spotted eagle rays

CLASS 3: OSTEICHTHYES — BONY FISH

Most of the world's fish belong to this class. They differ from the lampreys and sharks mainly by having a skeleton made of bone, and a single gill opening overlain by a gill cover on each side of the head (Figure 8.6). Many species also have scales although others have lost them through evolution. There is a great diversity in appearance.

Figure 8.7 shows fifteen families of bony fish that are likely to be seen.

Many types of deep sea fish swim to the surface at night. They are easily recognised because they have luminescent organs (photophores; Chapter 3, Figure 3.2) like little silver buttons on their bodies (Colour Plate 46). Families 1 and 2 below are regularly reported by seafarers.

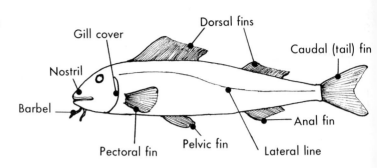

Figure 8.6: A generalised drawing of a bony fish

Dorsal fins

Gill cover

Caudal (tail) fin

Nostril

Barbel

Anal fin

Pectoral fin

Pelvic fin

Lateral line

Family 1:
Myctophids —
the lantern fish

Tiny golden orange/yellow lights darting about in the beam of a searchlight shone on the sea's surface at night are probably reflections from the eyes of lantern fish. These are small fish, about ten centimetres long, that migrate to the surface at night in huge shoals to feed on small planktonic shrimp-like krill. They are blackish silvery fish with large eyes and fragile scales. The photophores are in rows down each side of the belly (Colour Plate 46). Correct identification of the species depends to a large extent on the arrangement of these photophores. There are many species found all over the world, and sometimes they will be found on the deck after a night of heavy seas.

Family 2:

the hatchet fish

These small fish grow to seven centimetres long and are found in the deep waters of the world. One group of species — *Argyropelecus* spp. have been found several times recently on the decks of vessel steaming through the Bay of Biscay (Figure 3.1).

Family 3:
Exocoetids —
the flying fish

In flying fish the front (pectoral) fins and, in some species, the rear (pelvic) fins, are greatly enlarged and resemble wings. The lower lobe of the tail fin is also elongated to give an extra push on take-off (Colour Plate 47). Many different species can be seen in most tropical, sub-tropical and warm temperate areas of the world (Map 33). Generally the oceanic species are two-winged and grow to about 20 centimetres, while the better 'fliers' are the larger and more coastal four-winged species that may grow to 40 centimetres (16 inches). Both species swim just below the surface with their 'wings' folded along the body. If danger threatens — a chasing fish or an approaching ship — they launch themselves out of the water and, after taxiing across the surface by a sculling action of the lower lobe of the tail they glide out of harm's way. The four-wingers may cover up to 300 metres (nearly 1,000 feet) over the water, although normal flights usually cover between 30 and 100 metres (100 to 330 feet). Usually they glide about one or two metres above the surface but they have been reported hitting a bridge-wing 11 metres (36 feet) above the sea. They are attracted to light and this is used to catch them in some parts of the world. It also partly explains their frequent occurrence on the decks of ships.

Flying fish spawn in floating rafts of seaweed and other surface debris, and some species use the *Sargassum* weed, building nets of tightly packed balls of weed. The young, after they hatch, resemble the floating pieces of weed.

Family 1: Myctophidae — lantern fish (p. 88)
Size: 5–10 centimetres

Family 2: Sternoptychidae —
hatchet fish (p. 88) Size: 5 centimetres

Family 3: Exocoetidae — flying fish (p. 88)
Size: up to 60 centimetres

Family 4: Belonidae — skippers, garfish (p. 90)
Size: up to 1 metre

Family 5: Sphyraenidae — barracuda (p. 90)
Size: up to 2 metres

Family 6: Carangidae — yellow tails,
pilot fish (p. 90) Size: up to 2 metres

Family 7: Nomeidae — man-of-war fish (p. 91)
Size: up to 20 centimetres

Family 8: Coryphaenidae
— dolphin fish (p. 91) Size: 2 metres

Family 9: Istiophoridae — martins,
sailfish (p. 91) Size: 2.5 metres

Family 10: Xiphidae — swordfish (p. 92)
Size: 3.5, rarely 5, metres

Family 11: Echeneidae — sucker fish
(remoras) (p. 92) Size: 65 centimetres

Family 12: Aridae — sea catfish (p. 92)
Size: 45 centimetres

Family 13: Molidae — sunfish (p. 92)
Size: 3 metres

Family 14: Antennaridae — frogfish (p. 93)
Size: 15 centimetres

Family 15: Balistidae — trigger fish (p. 93)
Size: 45 centimetres

*Figure 8.7: A pictorial key to
fifteen families of bony fish*

**Family 4: Belonids —
the needlefish or
garfish and skippers**

These long slim fish have thin jaws and are found throughout the world (Colour Plate 48).

*Scomberesox saurus —
the Atlantic skipper or
saury pike*

The saury grows to 50 centimetres (20 inches) and is an oceanic species found throughout the North Atlantic — usually in large shoals swimming just below the surface (Map 34). They are able to escape predators by leaping out of the water and skipping along the surface on their tails. They can be distinguished from garfish by having a series of small finlets just in front of the tail. A similar Pacific species, *Cololabis saira*, is found from Japan to California (Map 35).

*Belone belone —
the garfish*

The garfish is very like the saury but occurs closer inshore during the summer, and often in estuaries. It is frequently seen from ships at anchor, particularly at night when it feeds on shoals of smaller fish attracted by the ship's lights (Colour Plate 48).

Many surface fish like flying fish, garfish and sauries are well camouflaged, dark green or blue above, and white below. However, some young needle fish living among tropical islands like the Virgin Islands are straw coloured and they can hold themselves motionless in the water so that they look just like the floating pieces of weed and grass that are particularly common around these islands.

**Family 5:
Sphyraenids —
the barracuda**

Barracuda have an undeserved reputation for attacking bathers, but although they are inquisitive and will approach anyone swimming, the records of unprovoked attack are rare. The fish resembles a pike, with a long slim body which is dull to bright silver and a mouth with a protruding bottom jaw bearing many sharp teeth (Colour Plate 15). Several species occur in shallow waters around the world (Map 36), and a number of species of Pacific barracuda, *Sphyraena* spp., are found from Alaska south to Peru, and in Australian waters.

*Sphyraena barracuda —
the great barracuda*

This species may grow to two metres (six feet) long, though this is rare. It is common from Florida to Brazil and, although a common sport fish, its meat is often toxic and not recommended eating. A similar but smaller species, *Sphyraena guachancho*, called the guaguanche locally, occurs in the same area and is excellent eating.

*Sphyraena sphyraena —
the European
barracuda*

These barracuda grow to 50 centimetres (20 inches) and live in the Mediterranean, and the east Atlantic north to Biscay.

**Family 6: Carangids —
the yellow tails
and pilot fish**

This large family, which includes the scads or horse-mackerels and the jacks, are mostly fast-swimming surface ocean fish. They are voracious hunters, widely distributed in tropical and temperate waters. Many, especially the young and the smaller species, have the habit of sheltering around large floating objects, and close to animals such as sharks, turtles and jellyfish (Colour Plate 32). generally these fish have flattened bodies well developed fins and a narrow tail stalk.

Frequently seen in shoals in the Mediterranean and southern Atlantic, they are unmistakable with their brilliant yellow colour. A closely related species, *Seriola dorsalis*, the California yellowtail, occurs off the Pacific coast of North America (Map 37).

Seriola dumerili — the Atlantic yellowtail or amberjack

This species is usually associated with sharks, mantas, turtles, boats and drifting debris (Colour Plate 16). It was originally thought that they ate parasites from their host, or even helped themselves to titbits after the host had eaten. But it is more likely that the fish are simply getting shelter and protection from larger predators. They grow to 35 centimetres (14 inches), but occasionally up to 70 centimetres (27 inches). The young often congregate in groups beneath the larger jellyfish.

Naucrates ductor — the pilot fish

There is only one species in this family — the man-of-war or bluebottle fish, *Nomeus gronovi*. The fish is commonly found sheltering under the Portuguese man-of-war, *Physalia* (Chapter 4).

Family 7: Nomeids — the man-of-war fish

There are a couple of species of dolphin fish (not to be confused with the true dolphin, which is a mammal) that are cosmopolitan in tropical and sub-tropical waters (Map 38). They are spectacular looking fish, having a fin extending from the steep forehead to the deeply forked tail. The body is blue-green above, silvery below, with silver and gold iridescence (Colour Plate 49). They are fast swimmers, feeding on other fish and squid, and are often seen leaping out of the water in a frantic chase after flying fish. Dolphin fish usually live in shoals, often undertaking long breeding migrations, and are commonly seen using the shade under floating weed and debris.

Family 8: Coryphaenids — the dolphin fish

C. hippurus is possibly the commonest species. The male grows to almost two metres (six feet) and as a mature adult has the characteristic steep forehead. The female is slightly smaller, with a more rounded forehead. Breeding usually occurs in summer and the fish is popular game and good eating. Colour Plate 49 shows the pompano dolphin, *C. equisalis* which is less common than than the dorado.

Coryphaena hippurus — the dolphin fish or dorado

Unmistakable fish with the upper jaw prolonged into a long cylindrical spear, a fin running down the back, and a sharply forked tail (Colour Plate 17). They grow between two and two and a half metres, are strong swimmers, and are usually brightly coloured — silvery underneath and metallic blue above. Some species have bands around the body.

Family 9: Istiophorids — the marlins, spearfish and sailfish

The sailfish has a very large blue dorsal fin — hence the common name. The Atlantic sailfish occurs from Cape Hatteras on the United States coast, south to Brazil, and rarely towards the European coast: the latter, however, are lone stragglers as the species is not normally migratory. The Pacific sailfish is found from southern California to Equador (Map 39).

Istiophorus spp. — the sailfish

Tetrapterus spp. — the marlins or spearfish	These species do not have such a large fin, although it does extend from the head to at least two-thirds the length of the body. *Tetrapterus belone* is a Mediterranean species: it is two metres long and has an unbanded body and a short spear. But *Tetrapterus albidus*, the white marlin, grows to two and a half metres. It is more greenish above, changing abruptly to silvery white with a white belly. It is an Atlantic species, but rarely seen around southern Britain (Map 40).

There are several Pacific marlins including the black marlin, *Makaira indica*, which is known from Mexico to Peru, and the blue marlin *Makaira nigricans*, which comes from Hawaii and is the largest of the marlins.

Family 10: Xiphids — the swordfish	The swordfish, *Xiphias gladius*, is the only species in this family and it may be confused with the marlins but the upper jaw is flat, not cylindrical and the dorsal fin is sickle-shaped and does not extend down the body (Colour Plate 18). They grow to almost three and a half metres and the body is dark grey, dark blue or brownish red, lighter on the sides with touches of bronze and a white-grey belly. They occur singly throughout the Mediterranean area and all tropical and temperate seas (Map 41).

Contrary to popular mythology, marlins and swordfish do not spear their food but rather lash out among shoals of fish then return to eat the ones that are dead and injured.

Family 11: Echeneidids — shark suckers or remoras	The remora or shark sucker has a ridged disc on top of the head which resembles the sole of a boot. It is a fin modified as a sucker for attaching to sharks, manta rays, turtles, large fish and even boats (Colour Plate 19). Remoras do not suck body fluids but feed mostly on parasites growing on their host. They first attach themselves while only three or four centimetres long — often to another remora at first but then changing hosts as they grow. Some remoras are specific to certain groups of animals; for instance *Remora brachyptera*, which is a dark brown fish, is usually attached to swordfish, sailfish or marlins; *Remora albescens* is a dull grey to cream colour and attaches to manta rays.
Remora remora — the common remora or shark sucker	The adult grows to 64 centimetres (25 inches) and is brown or greyish brown. They are found in the Mediterranean and all warm seas, but occasionally also in temperate waters.
Echeneis naucrates	*Echeneis* is another remora which is found only in warmer oceans. It has a distinct black stripe running along each side of the body.
Family 12: Arids — the sea catfish	Perhaps the most common is the gaftops'l catfish, *Bagre marinus*, usually found scavenging on sandy sea bottoms, especially in anchorages, where it is easily caught with a rod and line (Colour Plate 50). *A word of warning*' the front ray of the dorsal fin has an extremely sharp spine which, although not poisonous, can inflict a painful wound.
Family 13: Molids — the sunfish	Sunfish are found in the open oceans of tropical and temperate areas. They are large, disc-shaped fish with a leathery skin; they

have no scales, and the tail-fin is a frill running down the blunt rear end.

This spectacular fish is an intense dark grey or violet colour with silvery sides and belly. The body grows to 80 centimetres (31 inches) and is more elongated than the true sunfish. The fish has a vertical slit-shaped mouth. This species is rare and very little is known about its habits.

Ranzania laevis — the truncated sunfish

This giant grows to three metres (ten feet) in diameter and has a grey-brown back with lighter sides and belly. The mouth is very small, with a single beak-like tooth in each jaw (Colour Plate 20). Sunfish are probably deep-sea fish; when they are seen on the surface they are usually swimming feebly on their sides, suggesting that they are dying. They eat small jellyfish, sea squirts and sea gooseberries, but little else is known.

Mola mola — the sunfish

There are many species of frogfish although only one species, the *Sargassum* fish, *Histrio histrio*, is likely to be found in the open sea (Colour Plate 21). This species is widespread in the warmer areas of the Pacific, Atlantic and Indian oceans, but it is usually found among the *Sargassum* weed (Chapter 11, Map 160). The fish is extremely well camouflaged with a mottled yellow, brown and white body and ragged fins which resemble 'leaves'. The fish can climb among the weed using its arm-like pectoral fins, but it can also move by using jets of water from the round gill openings behind each 'arm'. The animal feeds on smaller fish which are sucked into the mouth in an instantaneous gulp. It grows to 10 centimetres but it can suck water in to pump itself up to dissuade any would-be attacker.

Family 14: Antennarids — the frogfish

These fish get their name from the interlocking device on the three spines in the first dorsal fin. The first spine can be locked in the upright position; to fold it away the second and third spines must be depressed. The body is flattened sideways; the eyes are set well back, and the front part of the head is hard to protect the fish from the spines of the sea urchins they eat. The skin is extremely tough. There are several species, the largest of which is the queen trigger fish, **Balistes** spp. (Colour Plate 51). This is easily distinguished by the trailing edges of the fins. Another species, *Melichthys niger*, the black fish, is extremely abundant around Ascension Island in the South Atlantic. It is a voracious predator and has been likened to the notorious freshwater piranha of South America, but the two species are not related. All trigger fish swim with undulating movements of the fins.

Family 15: Balistids — the trigger fish

9. The Reptiles

Turtles and Sea Snakes

The reptiles are probably one of the most notorious groups of animals. They are an ancient group, descended from the giant prehistoric dinosaurs, and are important because they were the first vertebrates to become fully adapted to living entirely on land: they do not return to water to breed like the amphibians (such as frogs and newts) do. Land-breeding was achieved mainly by the evolution of the hard-shelled egg which protects the embryo and keeps it bathed in fluid.

The evolutionary success of the reptiles can be measured by their world-wide distribution, and by the wide range of habitats they have occupied. The majority, including crocodiles, alligators, most lizards and snakes, tortoises and some turtles, live on land, in trees or in swamps, rivers and lakes; the remainder — large turtles, some snakes and one species of lizard — live in the sea.

CLASSIFICATION

Figure 9.1: Classification of marine reptiles

There are over 6,000 species of reptile but only 56 are found in the marine world. These include a marine lizard, about five species of turtles, and 50 sea snakes.

SUBPHYLUM:	CLASS:	ORDER:	FAMILY:
Vertebrata	Reptilia	1. Chelonia (turtles)	1. Chelonidae — green, hawksbill, loggerhead (p. 95)
			2. Dermochelydae — leatherback (p. 97)
		2. Squamata (snakes, lizards)	3. Iguanidae — the marine iguana (p. 97)
			4. Hydrophiidae — sea snakes (p. 97)

ORDER 1: CHELONIA — THE TURTLES

Turtles are easily recognisable animals. They resemble tortoises but the limbs have evolved into paddles and the head cannot be withdrawn into the shell. Like tortoises they have no teeth but possess a horny beak, although some species may have tooth-like serrations along one or both jaws. Generally, the male is slightly larger than the female: he has a longer tail, and the shell tapers more towards the rear.

Turtles are often seen many miles from land, basking in the sun or swimming leisurely along an instinctive course, pausing only to look with curiosity at a passing vessel before diving out of harm's way.

Not much is known about the marine turtles because of the inherent difficulty of following the animals in the wild. They start life as an egg buried about 60 centimetres (25 inches) down in a sandy pit above the high-water mark on a tropical beach. After hatching, the turtle is between five and ten centimetres long and, with its hundred brothers and sisters, it pushes upwards to lie just below the surface of the sand, waiting for nightfall. Then, under cover of darkness, they scramble out and run a gauntlet between hungry ghost crabs (Chapter 6) and seabirds until the fastest ones reach the relative safety of the sea. The young turtle is well camouflaged, its back is dark to match the sea; the underside is light to blend with the sky. And so the turtle virtually 'disappears'.

The next few years are spent wandering the oceans, in some cases for thousands of miles. Little is known about this wandering phase but after about seven years (the exact time varies for different species), the turtles return to the shore to breed, many going back to the beach where they were born. The male and female mate just offshore and the female, like her mother seven years previously, pushes her way up the beach above the high-water mark where she digs a hole and lays a clutch of a hundred or so eggs. After covering the eggs and spreading loose sand over the area the female digs a false empty nest 50 metres (165 feet) further away to lead any predators — including man — away. Then, excreting salt in tears from her eyes, she pushes herself back to the sea (Colour Plate 53).

The female will return again the following night, and every night for about three weeks, until she has laid about 1,000 eggs. Some nights she is unsuccessful either because the sand crumbles or because the hole is obstructed by roots, but she will continue undaunted until her task is done. She will then spend the next three years at sea before returning to breed again.

A turtle's life is precarious. Of the 1,000 eggs laid, on average only one will return and breed. It is a sad fact that due to the hunting and polluting activities of man, all the marine turtles are in danger of becoming extinct.

These turtles have paddles with one or two claws, enlarged in the male and used for clasping the female's shell during mating. They are spread throughout the world but in many cases they are divided into two races, the Atlantic and the Indo-Pacific.

Family 1: The Chelonids

The green turtle has a smooth shell, up to a metre long — slightly smaller in the female — with radiating wavy or mottled darker markings or larger blotches of dark brown, and a whitish to light yellow underside (Colour Plate 53). The turtle's name comes from the colour of the fat, not the shell. The tail of the male projects about 20 centimetres beyond the shell. It is curved downwards and tipped with a heavy nail used in helping to grasp the female.

Chelonia mydas — the green turtle

Green turtles make extensive migrations but this may simply be due to the currents carrying the animals (Map 42). They are usually seen in shallower water because their main food is seaweed, although they will eat jellyfish, shellfish and crabs.

The Atlantic green turtle occurs from the shores of New England in the United States to about 38° south, including the Gulf of Mexico, in the west, and from the Mediterranean to the Cape of Good Hope in the east. This species has a predominantly brownish shell which is not as chunky and deep as the Pacific species. The Pacific green turtle occurs in the Indo-Pacific from the eastern shores of Africa to the southern Asiatic coastline and Australia, and from southern California to Chile at about 43° south. The green turtle has been hunted for years for its meat and eggs which are regarded as delicacies. Expanding demands has pushed the animal to the verge of extinction and today there are few safe breeding sites remaining.

Eretmochelys imbricata — the hawksbill or tortoiseshell turtle

This species has a sharp ridge of overlapping plates running down the middle of the shell. The ridge is not so obvious in older animals so these may be confused with the green turtle. However, the hawksbills are smaller, with a shell-length of up to 80 centimetres (31 inches) for males and 60 centimetres (24 inches) for females (Colour Plate 52).

The Pacific hawksbill is heart-shaped and the upper surfaces of the head and flippers are almost solid black. They are found in the Indo-Pacific from Japan to Madagascar in the west and from Baja California to Peru in the east.

The Atlantic hawksbill shell is more straight-sided, with a narrower tapering behind, and the upper surfaces of the head and flippers are less black. They occur in the Gulf of Mexico and the Caribbean Sea and off the western coast of Africa around Morocco, (Map 43).

Like the green turtles, they mainly eat seaweed and hence are usually found in shallower water, especially in small mangrove-bordered bays, lagoons and estuaries. They will eat shellfish, jellyfish and animals such as *Physalia*, the Portuguese man-of-war, closing their eyes to avoid the harmful stinging tentacles. Hawksbill turtles tend to have excessive encrustations of barnacles and other animals on their shells.

The hawksbill meat is not good eating although the eggs are a delicacy. The main economic value is in the shell — 'tortoise shell' — which, in the past, was burnt off the live animal with a slow fire. They were then returned to the water where most would die from their wounds or from attacks by sharks and other fish.

Caretta caretta — the loggerhead turtle

Loggerheads are easily recognised by their reddish-brown elongated shell and large broad head. They grow to about a metre (three feet) long by about 70 centimetres wide and are great wanderers, often seen floating in open water many miles from land. They feed mainly on jellyfish, crabs, shellfish and fish.

The Atlantic population extends from Nova Scotia to Argentina, and from the British Isles to central West Africa. The Indo-Pacific population occurs from southern California and Chile to the Cape of Good Hope and the east coast of Africa (Map 44).

The eggs and meat of loggerheads are prized for food, but oil has also been extracted for use as varnish.

Lepidochelys spp. — the Ridley turtles

The Atlantic and Indo-Pacific populations of Ridleys (Map 45) are distinctly separate populations, and are often considered separate species. The Atlantic species, *Lepidochelys kempi*, occurs from Massachusetts to the Gulf of Mexico and is occasionally carried in the Gulf Stream to Britain and the Azores. It is the smallest of the Atlantic turtles, growing to between 60 and 70 centimetres, with a grey, heart-shaped to almost circular shell. They mostly eat crabs.

The Pacific Ridley, *Lepidochelys olivacea*, occurs as far north as southern Japan and Baja California, and south to Chile. The broad, flat-topped shell is a uniform olive colour, almost circular, and up to 70 centimetres in diameter. This species, although mainly vegetarian, will eat shellfish and sea urchins.

The Ridleys have been hunted for their eggs, and sometimes for use in veneering and inlaying.

Family 2: Dermochelyds

The leatherback is the only species in this family although there are Atlantic and Pacific populations (Map 46). Leatherbacks are the largest of all turtles, growing to two metres (six feet) long with seven ridges running down the back and five down the underside. The front flippers are much larger than the back, and the animals are strong swimmers — often found well out to sea. The limbs have no claws, and the shell is composed of a mosaic of small bony plates covered in a dark leathery skin. This is the only turtle that makes a noise, rather like a grunt or hollow bellow.

Dermochelys coriacea — the leatherback turtle or luth

The Atlantic population occurs from the Gulf of Mexico south to the Plata in Argentina and north to Nova Scotia, and from the British Isles to the Cape of Good Hope. The Indo-Pacific population is found from Japan to southern Australia and the Cape of Good Hope, and from British Columbia to New Zealand and Chile. They feed on jellyfish, crabs, shrimps and fish.

This turtle was hunted mainly for the oil yielded by the skin, and for the shell which was used in varnishing. The eggs are eaten wherever they are available.

ORDER 2: SQUAMATA — THE SNAKES AND LIZARDS

There is only one species of marine lizard — *Amblyrhynchus*, the marine iguana, which is found among the Galapagos Islands in the Pacific ocean.

Family 3: Iguanids

Sea snakes are among the most poisonous of all snakes having a pair of large erectile fangs at the front of the mouth (Figure 9.2). They are widely distributed in tropical seas from the Persian Gulf and the east African coast to Central America (Map 47), but they are not found in the Atlantic. Some live on the sea shore and make excursions out to sea; others remain entirely at sea.

Family 4: Hydrophiids — the sea snakes

Sea snakes are well adapted to a marine life. They have a flattened tail for swimming — sometimes the whole body is flattened sideways — and the small eyes and nostrils are on top of the head. Most species do not lay eggs but give birth in the sea. There are about 50 species.

Figure 9.2: A generalised drawing of the head of a sea snake

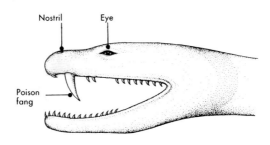

10. The Mammals
Whales, Dolphins, Porpoises and Seals

Many seafaring stories tell of dolphins leading ships through dangerous channels or saving seamen who have fallen overboard. And perhaps no sight is as popular at sea as that of a group of dolphins riding the bow wave. The marine mammals range from the smallest seal of less than a metre (two or three feet) to the largest animal ever to have lived — the blue whale which may reach a length of 30 metres (100 feet). Mammals are warm-blooded, keeping their body at the same temperature regardless of changes outside, and they give birth to young that are fed with milk produced by mammary glands. Man is a mammal, and hence the whales, dolphins and seals represent our closest relatives in the sea.

It is believed that life originated in the sea some 3,000 million years ago, but it wasn't until about 350 million years ago that some animals started living on land. From these primitive organisms all living things evolved. Mammals developed on land, but the ancestors of the marine mammals probably lived in muddy regions near the sea. They may have obtained their food from the sea and gradually developed into marine creatures.

Through evolution the whales, dolphins and porpoises gradually lost their back limbs and developed a horizontal flipper or **fluke**; the front limbs were retained, and changed into flippers (Figure 10.2). The animals are streamlined, and all projections such as nipples and external ears have been enclosed in a fold of skin — or lost. Even the penis, which may be up to three metres long in the large whales, is held back in a fold of the skin by muscles attached to what remains of the back limbs. Evolution has made these mammals into very efficient swimmers; perhaps the only drawback is that they are air-breathing. In whales and dolphins the 'nose' has developed far back on the head, so it is usually the first thing to break the surface. They can also breathe out and in very quickly; for example, the fin whale exchanges over 3,000 times the air volume of the human being in less than two seconds.

CLASSIFICATION

There are three groups of truly marine animals, but a fourth group, the sea otters, spend much of their time in shallow seas (Figure 10.1).

In the descriptions below, the measurements given are usually the maximum attained by the individual species, and there is usually a great deal of variation.

ORDER 1: THE CETACEANS — WHALES, DOLPHINS AND PORPOISES

Cetaceans are the most advanced marine mammals. They have lost all connection with the land — eating, mating and giving birth entirely in the sea. Most cetaceans use sound to communicate, and as sound travels four times faster through water than through the air, it

Figure 10.1: Classification of
marine mammals

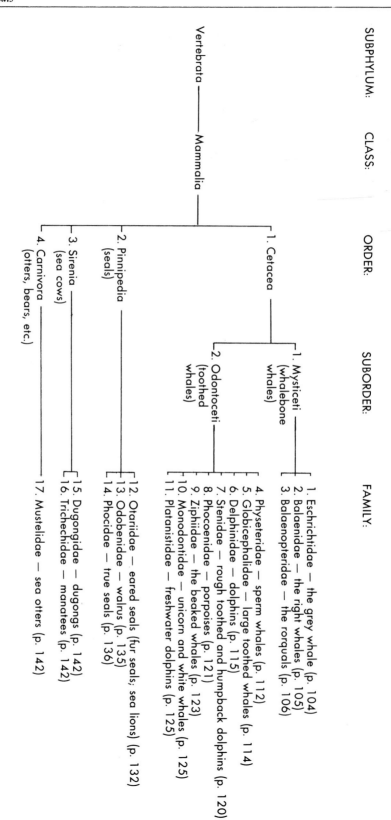

is possible that some animals communicate over large distances. Sound is also used as an aid to navigation. In a similar way to bats, a sound is emitted and the echo received. This is **echo-location** — the echo tells the animal about its surroundings and, in some cases, whether an object is soft or hard. Unlike bats' sound, which is usually supersonic, some whale sounds can be heard by man. Recent research shows that sound may be used to catch food, as some dolphins can produce a noise of such an intensity that it will stun or even kill fish at close range.

Figure 10.2: Generalised drawing of the two types of whale (cetaceans)

Whalebone whale — mysticete

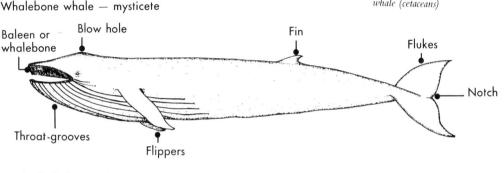

Toothed whale — odontocete

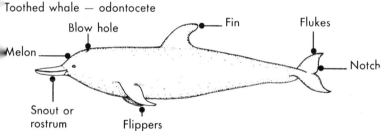

Many of the larger species of whale undertake long migrations from cold waters to tropical areas and back. Their breeding cycle is related to this migration pattern. The whales meet to breed in the warmer areas and then travel north or south. They return the following year to give birth, hence the pregnancy or gestation period is usually around eleven or twelve months. During the time in the tropics the mother is usually fasting. The young are fed on milk for up to the first twelve months.

Birth is probably one of the most difficult times. All species are born tail-first, and often there are other animals in attendance behaving like 'aunts' or 'midwives' to assist and protect the young and mother. As soon as the young is born it is pushed to the surface by the mother or one of the 'aunts' for its first breath. Then the calf swims very close to the mother. The instinctive behaviour of pushing the young to the surface will persist even if the calf is stillborn — a captive bottlenose dolphin was recorded holding its dead calf at the surface for nearly four hours. This helps to explain why some people have been saved from drowning by dolphins keeping them at the surface, and may explain why some cetaceans have been seen apparently playing with objects such as pieces of wood at the surface.

There are between 75 and 80 different species of whales, dolphins and porpoises: the exact number is debated because of their wide distribution. For instance, in some cases there are northern and southern varieties which may have evolved slight differences; some people divide these into two species but it is probably better to label them as subspecies.

Cetaceans are divided into two groups: those with between one and 200 teeth (Odontocetes) and those with whalebone or baleen (Mysticetes). The whalebone or baleen is a giant 'sieve' that resembles a system of brushes hanging from the roof of the mouth (Figure 10.3).

Figure 10.3: (a) A section through the jaw of a rorqual showing the baleen. (B) Baleen from three whalebone whales.

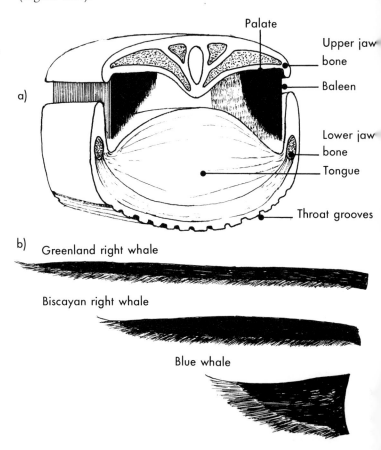

Palate

Upper jaw bone

Baleen

a)

Lower jaw bone

Tongue

Throat grooves

b) Greenland right whale

Biscayan right whale

Blue whale

Initially there may be difficulty in identifying all the cetaceans, partly because some species may only be seen at a distance, and then only briefly. However, each species of large whale produces a blow with a particular shape, in addition to which the surfacing and diving characteristics can help in identification of the species (Figure 10.4). The old whalers were adept at identifying whales at a distance but with practice, and using the outlines shown in Figure 10.4 and the notes following, a seafarer can quickly gain a similar proficiency. Some smaller species will approach ships and even ride the pressure wave in front of a bow. This helps in identification. Other animals will avoid ships, and this also is a clue to the species. The best way of identifying the animals is to watch them for as long as possible, make

Figure 10.4: The surfacing and diving characteristics of some large whales

1. *Balaenoptera musculus*
— the blue whale (p. 107).

Head is broad and flat; the fin is about 30 centimetres high; the flukes are seldom raised on the dive.

2. *Balaenoptera physalus*
— the fin whale (p. 107).

White right lower jaw; the fin is about 60 centimetres high; the flukes are never raised on the dive.

3. *Balaenoptera borealis*
— the sei whale (p. 108).

The fin is about 60 centimetres and seen with the blow; the dive is shallow — the animal appears to sink from sight.

4. *Balaenoptera edeni*
— Bryde's whale (p. 108).

The fin is about 45 centimetres tall and curved; the flukes are not raised on the dive.

5. *Megaptera novaeangliae*
— the humpback whale (p. 106).

Extra long flippers with white underside; the flukes white on the underside; head flat and knobbly.

6. *Balaena* spp.
— the right whales (p. 105).

Double blow; no fin; flukes raised on the dive.

7. *Eschrichtius robustus*
— California grey whale (p. 104).

Distinct mottled colour; flukes raised on the dive.

8. *Physeter macrocephalus*
— the sperm whale (p. 112).

Large box-like head; forward projecting blow; flukes are raised on a deep dive.

brief notes about their numbers and behaviour, sketch the outline, particularly the shape of the fin, and note the size, shape and direction of any blow. Then refer to the descriptions. Remember the ship's position is important. Information on cetaceans is extremely valuable to biologists such as those running the International Dolphin Watch whose address is given in the appendix. This, and other organisations, will be able to supply more information.

1. THE MYSTICETES — BALEEN OR WHALEBONE WHALES

There are three families of baleen whales: the Eschrichtidae, with only one species — the California grey whale; the Balaenidae or right whales, and the Balaenopteridae or rorquals. These whales feed by engulfing water and the food — small fish and krill — and then squeezing out the water through the baleen plates, leaving the food trapped inside.

The baleen whales are all large whales and are sometimes referred to as the great whales. Generally they grow to over 13 metres (43 feet) long, with the female being larger than the male.

Family 1:
Eschrichtidae

The one species in the family is the California grey whale *Eschrichtius robustus*. It is the most primitive baleen whale, with characteristics of both right whales and rorquals. Today, grey whales are only found in the coastal waters of the North Pacific.

Adults may reach 17 metres (56 feet) in length. They have a slender body and a small head, about one-fifth of the body length. The skin is dark muddy grey with many lighter patches, and it is always infested with clusters of barnacles and whale lice (Chapter 2, Figure 2.16). There is no fin, but a line of eight or ten small humps or knuckles extends down the tail. The flukes are curved slightly and are usually raised above the sea when the animal dives.

The conservation of grey whales is a success story. Because this whale has a regular coastal migration it was vulnerable to whale hunters from shore and sea. The estimated Pacific population in the mid-nineteenth century was about 30,000, but it was reduced to non-profitability by the end of the century. A resumption of whaling in the 1920s and 1930s brought the animals to the point of extinction, and they were protected by law in 1937. From that date the numbers have increased until, in 1971, the eastern Pacific species became stable at 11,000. The western Pacific (Korean) population did not fare so well and is now probably extinct.

The annual migration of the grey whale is the longest undertaken by any whale (Map 48). The summer feeding grounds are in the Bering and Chukchi Seas off Alaska. From here they travel south to the shallow lagoons of Baja California where they calve in January and February. The northbound journey begins about March to April, the newly pregnant females being the first to leave followed by the new mothers and calves. Because their arrival off various points *en route* is fairly predictable these whales have become a tourist attraction for the west coast of America and Canada.

Grey whales are very active. They regularly leap out of the

water and crash back — a behaviour known as breaching. This serves to remove the irritating barnacles and whale lice. These whales will occasionally 'stand' with just their heads poking out of the sea, apparently looking around — this is called spy-hopping.

These whales are slow swimmers, attaining about three to four knots, and they float when they are killed — hence they were the 'right' whales to hunt in the early whaling days. There are three species, all with high arched jaws enclosing a large mouth — the main characteristic of this group.

Family 2: Balaenidae — the right whales

This whale was the basis of the eighteenth century whaling industry, when it was described as the common whale. Now it is one of the rarest (Map 49). The bowhead grows to 20 metres (66 feet) long and has a maximum circumference of 9 to 12 metres (30 to 40 feet). The head occupies one-third of the length of the body. In 1820, William Scoresby, who was perhaps the first academic whaling expert, as well as being a very successful whaling captain, wrote that the head "... may, when the mouth is open, present a cavity capable of containing a ship's jolly-boat full of men, being 6 or 8 feet wide, 10 or 12 feet high and 15 or 16 feet long". The baleen is three to three and a half metres (ten to twelve feet) long in the centre of the upper jaw (Figure 10.3).

Balaena mysticetus — the Greenland right whale, bowhead, or great polar whale

The flippers are broad, there is no fin, and the back is completely smooth; the flukes are broad and curved. The skin is velvety black in the adult, but the young are lighter. The lower jaw is white along one-third of its length. From the side, the whale has a distinctive double-humped appearance, the first hump is the curve of the head; the second is the broad expanse of the back. On sounding, the tail flukes are shown above the sea and the whale returns to the surface near the original position. A characteristic V-shaped blow is produced by the twin blowholes (Figure 10.4).

This right whale and its almost identical southern hemisphere relative *Balaena glacialis australis*, may grow to about 16 metres (52 feet) long. They are similar in shape to the bowhead but with a smaller head, which has greyish-white horny outgrowths at the front of both jaws. These are **callosities** and they form an individual pattern which distinguishes this species from any other. The skin is generally black or brownish, sometimes with an irregular area of white under the chin or belly. When diving they throw their flukes high above the sea, showing the broad outline, pointed tips, and deep central notch.

Balaena glacialis — the black right whale, Biscayan right whale, or nordcapper

The northern right whale is found in the Atlantic from the Denmark Strait to the Caribbean and from the Norwegian Sea to the Canary Isles. In the Pacific they can be seen in a broad arc between the Sea of Japan, through the Aleutians and down the coast of California (Map 50). A general migratory movement southward occurs at the end of Autumn, with a northward return in the Spring.

The southern right whale is found throughout the southern

ocean with seasonal concentrations around Western Australia, New Zealand, Chatham Island, southern Chile and southwest Africa (Map 51).

Caperea marginata —
the pygmy right whale

This is a smaller whale — only about 6.5 metres (20 feet) long, and more slender so it may be mistaken for a small rorqual, but a close look at the head shows the characteristic high arched jaws. It has narrow flippers, a curved fin two-thirds of the way along the back, and the skin is smooth, with no growths or callosities. Few animals have been examined alive, but it has been described as having a dark grey back that merges into a paler underside. The upper half of the lower jaw is dark and a dark band runs between the eye and the flippers.

Pygmy right whales are only found in the southern waters, generally to the north of the Antarctic convergence (Map 52). They have been recorded from South Africa, South Australia, Tasmania, New Zealand, the Falkland Islands and the French Antarctic Islands.

Family 3:
Balaenopteridae —
the rorquals

The name 'rorqual' is derived from the Norse word 'rørhval' which refers to the grooves or folds on the underside of the throat which are characteristic of these whales. These throat grooves or pleats enable the mouth to stretch and engulf a large volume of water when the animal is feeding, and by contracting force water out through the baleen plates. All rorquals have a fin about two-thirds of the way along the back, and usually a ridge running forward from the blowhole. They are fast swimmers and sink when they are killed. These facts protected the whales from the early hunters and it was not until the advent of fast steam catcher ships and the invention of the explosive harpoon in 1868 that these swift, powerful animals could be hunted.

Megaptera novaeangliae —
the humpback whale

Humpbacks grow to 16 metres (52 feet) and have a distinct shape. They are fairly robust behind the flippers, narrowing towards the tail. The flippers are long, almost one-third the length of the body; they are scalloped along the leading edge and often have barnacles attached (Colour Plate 22). The flukes are raised high in the air when the animal dives, and they are broad, curved, with a serrated trailing edge and a deep central notch. The skin is black and white on the throat, belly, flippers and flukes.

In these whales the ridge from the blowhole is indistinct but the top of the snout and the edge of the lower jaw is covered by a series of fleshy knobs about the size of a cricket ball. The tip of the lower jaw is usually covered with barnacles and whale lice. At the end of the polar season humpacks may carry as much as 450 kilograms (about half a ton) of barnacles. The fin is small and irregularly shaped, often with a step or hump faired into the bottom of the leading edge. The blow is small — about three metres (ten feet), and rather bushy.

Humpbacks are found in northern and southern oceans with several distinct populations (Map 53). Annual migrations follow precise routes from summer feeding areas in the higher latitudes to winter breeding grounds in bays and islands of tropical waters. Until 1973 the numbers were severely depleted, but since then there has been an increasing number of observations off the northwest coast of North America, the Queensland coast and the southern ocean, so perhaps some recovery is taking place.

The blue whale is the largest animal alive today and the largest animal ever to have lived, bigger than any of the prehistoric dinosaurs. It grows to a length of 30 metres (100 feet) and weighs up to 100,000 kilograms (100 tons). A new born blue whale is about 7 metres (23 feet) long and over 2,000 kilograms (two tons) weight — and this will double in the first week.

Balaenoptera musculus — the blue whale or sulphur bottom

These are long slender whales with a small curved or sometimes triangular fin about 30 centimetres (12 inches) high. The head has a broad flat upper jaw with a central ridge, and the skin is bluish grey with dull white spots and mottling. Along the underside, especially after a long exposure in cold waters, there is often a covering of yellow plantplankton (diatoms — *see* Chapter 2) — hence the name sulphur bottom.

Blue whales feed mainly on the small shrimp-like crustaceans or krill (Chapter 2; Colour Plate 20) that are abundant in polar waters. An average blue whale will consume about four tons of krill a day, but this amount is restricted to a period of about 120 days before the ice closes over the feeding grounds, then the whales fast until the following spring.

Blue whales are not easy to identify and they may be confused with fin whales (see below), but the blue whale fin is smaller and they occasionally raise their flukes on diving. Blues are usually seen singly or in pairs.

These whales are found world-wide but they are rare due to intensive hunting. They range from the ice edge to the warmer temperate seas, sometimes into tropical waters (Map 54), with a general migration in the autumn months away from the ice to the warmer waters to calve. They have been seen in both shallow and deep water areas.

A subspecies — the pygmy blue whale *Balaenoptera musculus brevicauda* has been described from the area of the Kerguelen and Crozet Islands in the southern Indian Ocean. This whale grows to about 23 metres (75 feet) and is silvery blue with numerous spots and scars.

The fin whale is the second largest whale, growing to 27 metres (88 feet) in the southern oceans though sometimes smaller in the north. They are long and slender with a ridge running down the snout. Behind the curved fin the body forms a ridge — hence the name razorback. Fin whales are usually dark grey with a brownish tinge, but the underside (including the flippers and flukes) and the right side of the lower jaw is white; the left side is dark. Often a light, broad chevron is seen across the back behind the flippers.

Balaenoptera physalus — the fin whale, finner, finback, razorback, or common rorqual

Fin whales can swim at up to twenty knots, but eight knots is more usual. When surfacing the whale will come up at a steep angle and blow before the fin is in view (Colour Plate 54). The body makes a high arch on diving and the flukes are not shown.

They feed on a wide range of animals including krill and small fish such as herring and capelin.

These whales occur in similar areas to the blues, but they may not go as close to the ice (Map 55). A spring and autumn migration takes place, usually far offshore, but some have been seen in the western Mediterranean in the summer months. Calving takes place in summer.

Balaenoptera borealis — the sei whale (pronounced 'say')

Sei whales are similar to the blue and fin whales, and they are found world-wide. The southern animals grow to 17 metres (56 feet) but the northern whales are slightly smaller. They have small flippers and a fin about 60 centimetres (24 inches) tall which is sharply curved on both edges and further forward than in the fin whale. They are dark steel grey all over except on the underside of the forward part where they are greyish white, and there are many circular scars and blotches, possibly due to parasites.

Sei whales do not dive deep but skim a range of animals living just below the surface. They prefer krill and other crustaceans in the Antarctic but will also take small fish such as sardines, anchovies and young mackerel. Because of their feeding pattern the dive profile is characteristic. The whale surfaces at a shallow angle so that the blow and fin are seen in view together, and the dive is so shallow that the whale seems simply to sink from sight.

The whales occur off the Finmark coast where their arrival coincides with that of the sei or coalfish from which they derive their name. They also occur off southern Alaska and western Canada from May to August and off Kamchatka, USSR, from June to September; other groups live throughout the southern ocean (Map 56). Their winter distribution is less well known, although they appear to move towards tropical and sub-tropical seas at the end of summer and return the following spring. Calving takes place in the warmer waters.

Balaenoptera edeni — Bryde's whale (pronounced 'brewder') or tropical whale

Bryde's whale resembles the sei and fin whales but it is more slender. They grow to about 18 metres (59 feet) and have a curved, pointed fin about 46 centimetres (18 inches) high which often has a ragged trailing edge. Three ridges run forward from the blowhole and, as the whales are not shy and will often approach a vessel, these ridges may be seen easily. Two subspecies have been described, one living offshore and a slightly smaller one inshore. But they all have a dark grey back, a slightly lighter underside, and there is often a light flank patch just forward of the fin. They have many white circular scars, especially on the rear, which give the animal a galvanised look. The whale takes its name from J. Bryde, a Norwegian consul to South Africa who built the first whaling factory at Durban in 1909.

The diving characteristics are similar to the fin whale; the snout emerges at an angle and the body makes a high arch. The flukes are not raised. They feed on krill, small fish like anchovies, and occasionally baby squid.

These whales are not found in cold waters — hence the name

tropical whale is more definitive. They occur off South Africa and southern Japan, off Mexico and the Hawiian Islands, in the Caribbean and tropical north Atlantic (Map 57). Other groups occur near eastern Australia and the Northland Peninsula of New Zealand.

Minke whales are the smallest rorquals, ranging from 7 to 9 metres (23 to 30 feet) in length. There are large differences in size and colour and many believe there are three subspecies: *Balaenoptera acutorostrata* in the North Atlantic, *Balaenoptera acutorostrata davidsoni* in the North Pacific, and *Balaenoptera acutorostrata bonaerensis* in the southern hemisphere. A fourth subspecies has been suggested from Sri Lanka but this is not generally accepted.

Balaenophtera acutorostrata — the minkle whale, little piked whale, lesser rorqual or sharp headed finner

They all have a small head which is almost triangular with a sharp point at the front. The middle of the body is robust and the fin is tall, erect and curved at both edges. The back is dark grey to almost black and the underside is white to just short of the tail. A white band extends up the sides just behind the flippers, but there are variations in the amount of white on the flanks and flippers — some having a distinct band, others a light chevron like the fin whale. There is a characteristic white spot on the flipper of the northern whales.

North Atlantic minke whales feed mainly on fish like capelin, herring and sand lance; in the North Pacific krill and anchovies supplement the diet, and the southern species feed mainly on krill.

Minke whales are found in polar, temperate and tropical waters (Map 58). They enter small bays in polar regions and often get trapped in ice and drown. Elsewhere they will enter river mouths and strandings may occur. These whales are unafraid of ships and will often approach closely, in which event the characteristic white spot on the flipper can be seen to give easy identification.

2. THE ODONTOCETES — TOOTHED WHALES

The toothed whales are the largest group with about 65 species including 5 living in fresh water. The teeth are simple pegs with a root and a conical crown. They have one permanent dentition — they don't have any 'milk' teeth. All toothed whales have a single blowhole, which may be a plug or slit usually slightly to the left of the mid-line, and most usually have a fin. Males are generally larger than females.

The classification of toothed whales is constantly being revised as more information comes to light. Some workers create new families and species while others include animals in present groups. For instance, all the dolphin-like mammals have been either put into one family or divided into four families. I have used the separate families in these notes because they conveniently split the different types. But the differences may not be recognised by other authors as sufficient justification for separate families. Confusion may also arise because the slight differences in shape and colour have led people to describe subspecies. But more information from whale watchers will help to clear up some of these classification problems.

A guide to identifying different species is provided when the animals bowride. Not only does this enable the animal to be seen clearly, but the fact that it is bowriding is itself a clue. But it must be remembered that whether or not a species will come to the bow is not a hard and fast rule and may depend on the engine vibration, the speed of the vessel, or some other factor such as whether or not the animals were feeding. Some species that usually bowride may totally ignore a vessel and vice versa.

Figure 10.5: Outline drawings of some of the toothed whales (odontocetes)

Family 4: Physeteridae — sperm whales
Kogia breviceps — pygmy sperm whale
(p. 114). Length: 3.5 metres

Kogia simus — dwarf sperm whale
(p. 114). Length: 2.75 metres

Family 5: Globicephalidae
Orcinus orca — killer whale (p. 114).
Length: 10 metres
*

Pseudorca crassidens — false killer whale
(p. 114). Length: 5.5 metres
***

Feresa attenuata — pygmy killer whale
(p. 115). Length: 2.75 metres
?

Globicephala malaena — longfin pilot
whale (p. 115). Length: 6 metres
*

Delphinus delphis — common dolphin
(p. 116). Length: 2.5 metres
***

Stenella longirostris — spinner dolphin
(p. 116). Length: 2 metres
***

Stenella attenuata — bridled dolphin
(p. 116). Length: 2 metres
**

Stenella plagiodon — Atlantic spotted
dolphin (p. 117). Length: 2.5 metres
***

Stenella coeruleoalba — striped dolphin
(p. 117). Length: 2.5 metres
***

Stenella clymene — clymene dolphin
(p. 117). Length: 2 metres
? likely

Lagenorhynchus obscurus — dusky dolp
(p. 118). Length: 2 metres
***

Lagenorhynchus australis — Peale's
dolphin (p. 118). Length: 2 metres
***

Lagenorhynchus cruciger — hourglass
dolphin (p. 118). Length: 2 metres
***

Cephalorhynchus commersoni —
Commerson's dolphin (p. 119).
Length: 1.5 metres
**

Cephalorhynchus eutropia — black
dolphin (p. 119). Length: 1.5 metres
?

Cephalorhynchus heavisidi — Heavisid
dolphin (p. 119). Length: 1.4 metres
?

Globicephala macrorhyncha — shortfin pilot whale (p. 115). Length: 5.5 metres

Peponocephala electra — melonhead whale (p. 115). Length: 3 metres

Family 6: Delphinidae — the dolphins
Tursiops truncatus — bottlenose dolphin (p. 115). Length: 4 metres

Grampus griseus — Risso's dolphin (p. 120). Length: 4 metres

Orcella brevirostris — Irrawaddy dolphin (p. 120). Length: 2 metres

Lagendelphis hosei — Fraser's dolphin (p. 120). Length: 2.5 metres

Family 7: Stenidae
Steno bradensis — rough tooth dolphin (p. 120). Length: 2.75 metres

Sousa chinensis — Indopacific humpback dolphin (p. 121). Length: 3 metres

Lagenorhynchus albirostris — whitebeak dolphin (p. 117). Length: 3 metres

Lagenorhynchus acutus — Atlantic whiteside dolphin (p. 117). Length: 2.75 metres

Lagenorhynchus obliquidens — Pacific whiteside dolphin (p. 118). Length: 2.25 metres

Phocoena spinipinnis — Burnmeister's porpoise (p. 122). Length: 1.75 metres

Neophocana phocaenoides — finless porpoise (p. 122). Length: 1.75 metres

Family 9: Ziphiidae — beaked whales
Berardius arnuxi — southern giant bottlenose whale (p. 123). Length: 10 metres

Berardius bairdii — Baird's beaked whale (p. 123). Length: 13 metres

Hyperodon ampullatus — northern bottlenose whale (p. 123). Length: 10 metres

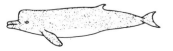

Cephalorhynchus hectori — Hector's dolphin (p. 119). Length: 1.8 metres

Lissodelphis borealis — northern right whale dolphin (p. 119). Length: 2.75 metres

Lissodelphis peroni — southern right whale dolphin (p. 119). Length: 2 metres

Mesoplodon grayi — scamperdown whale (p. 124). Length: 4 metres

Mesoplodon carlhubbsi — Hubb's beaked whale (p. 124). Length: 5 metres

Mesoplodon stejnegeri — sabretoothed whale (p. 123). Length: 6 metres

Mesoplodon ginkgodens — gingko toothed whale (p. 124). Length: ?

Mesoplodon mirus — True's beaked whale (p. 124). Length: 10 metres

Sousa teuszii — Atlantic humpback dolphin (p. 121). Length: 2 metres ? likely

Sotalia fluviatilis — tucuxi (p. 121). Length: 1.75 metres ?

Family 8: Phocoenidae — the porpoises Phocoena phocoena — common porpoise (p. 121). Length: 1.75 metres *

Phocoenoides dalli — Dall's porpoise (p. 122). Length: 2 metres * * *

Hyperodon planifrons — southern bottlenose whale (p. 123). Length: 9 metres

Tasmacetus shepherdii — Shepherd's beaked whale (p. 123). Length: 6 metres

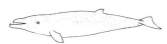

Ziphius cavirostris — Cuvier's whale (p. 124). Length: 6 metres

Phocoena dioptrica — spectacled porpoise (p. 122). Length: 5 metres ?

Mesoplodon europeus — Gulf stream beaked whale (p. 124). Length: 6.7 met

Mesoplodon densirostris — densebeak whale (p. 125). Length: 4.8 metres

Mesoplodon bidens — Sowerby's wh (p. 125). Length: 6 metres

Mesoplodon layardii — strap tooth w (p. 124). Length: 2 metres

Bowriding:
* * * — regularly seen bowriding
* * — often seen bowriding
* — Occasionally seen bowriding

? — no information
? likely — no information but may bowride

Family 4: Physeteridae

This is the sperm whale family and there are three species.

Physeter macrocephalus — the sperm whale

The sperm whale is the largest toothed whale; males reach 19 metres (62 feet) and females about 13 metres (43 feet). They are probaly the easiest whale to recognise, regardless of how much is seen of the animal. The whale's large box-like head i about one-third of the total length; in large males it ma overhang the relatively puny bottom jaw by up to one and a hal metres (five feet). Only the lower jaw has teeth. There is no fi but there is a distinctive rounded hump about two-thirds of th way along the back which is followed by a series of four or fiv smaller humps. The flukes, which are raised when the anima dives, are broad, triangular and smooth-edged, with a dee central notch.

The skin is dark grey or black, lighter on the belly, and wrinkled, especially at the rear. Older animals are heavil scarred (Colour Plate 23). Often there are white blotches aroun the mouth. But the main characteristic identifying the sperm whale is the blow, which is directed obliquely forward at abou 45 degrees (Figure 10.4). When viewed from the rear, th blow can be seen to come from the left side of the head. Th first exhalation after a deep-dive sounds like an explosion and can be heard at a distance.

The head contains the spermaceti organ. This is made up of tissue and a fine colourless oil which solidifies into a white wax in air. This was prized by the whale industry, providing candles, lubricants and polishes. The organ is believed to act as a buoyancy regulator during deep diving.

It is possible that water is taken into the nasal passage which may be up to five metres (15 feet) long and runs through the spermaceti (Figure 10.6). This action may cool the oil and increase its density so the head becomes heavier enabling the whale to achieve neutral buoyancy in deeper, colder waters. It may then be able to 'hang' quietly in the water listening for its prey, the giant squid. On its return to the surface extra blood may be circulated through the spermaceti to warm it. We may never be able to observe directly what a diving sperm whale does but the curious structure of the head is one of the most remarkable in the animal world.

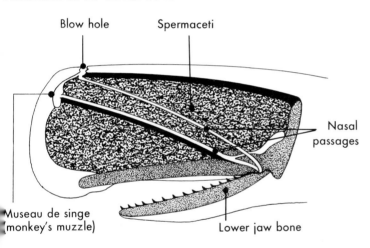

Blow hole   Spermaceti

Nasal passages

Museau de singe (monkey's muzzle)

Lower jaw bone

Figure 10.6: A section through the head of a sperm whale

Sperm whales descend to great depths to find the squid that form the greater part of their diet in most offshore areas. Round scar marks on the body are probably sucker marks from the tentacles of giant squid (Chapter 5). Fish are eaten especially when close inshore. The largest whales dive regularly to depths of over 1,200 metres (3,900 feet) and can remain down for up to an hour. One whale caught off South Africa spent an hour and 52 minutes underwater before it was caught. It is likely that even greater depths are attained. Sonic recordings have been made at over 2,000 metres (6,500 feet), and bottom-living sharks have been found in the stomachs of whales where the sea-bed is deeper than 3,000 metres (9,840 feet).

The whales take 30 or 40 breaths before starting a deep dive, that is they hyperventilate — something a human diver must never do. At the start of a deep dive the flukes are raised above the surface and the animal sounds, but returns very close to the same spot. This suggests that rather than hunt their food they lie in ambush. Giant squid have sense organs and could detect and avoid a swimming whale — they certainly manage to avoid nets.

It is probable that only by the whale lying still will a squid come close enough to be caught. The narrow bottom jaw may be an efficient 'snapping' device in water.

Sperm whales are found world-wide in temperate and tropical waters (Map 59). Only the large bulls will enter the polar waters. Moby Dick was probably a white or grey coloured sperm whale.

Kogia breviceps — the pygmy sperm whale

This whale grows to about 3.5 metres (11 to 12 feet). The body is plump with a small curved fin slightly behind the mid-point. The skin is a steel blue-grey colour on the back but lighter on the sides and belly. The head is squarish but the main distinguishing feature is a bracket mark like a fish gill in front of the flippers.

The pygmy sperm whale is found in warm temperate waters but it is not often seen, and information is scarce (Map 60).

Kogia simus — dwarf sperm whale

The dwarf sperm whale is similar to the pygmy whale but it is smaller — only about two and three-quarter metres (nine feet) long. The gill-like bracket mark is also present but the fin is in the middle, taller and dolphin-like. This species is mainly tropical and sub-tropical although not much is known (Map 61).

Family 5: Globicephalidae

This group of whales is sometimes included in the dolphin family.

Orcinus orca — the killer whale

The killer whale may grow to 10 metres (33 feet) and is one of the most easily recognised species. It has a rounded head with a large white patch above the eye. The body is black on the upper surface with a grey 'saddle' behind the fin; the underside is white, and a white flash reaches up and back along the sides. The fin of the male is large, about two metres (six feet) high, and straight although its weight may cause it to bend. The female's fin is smaller and curved slightly.

Killer whales feed mainly on fish, seals, penguins, and porpoises but they will combine to attack a large whale. They tear lumps from the lips, fins and bottom of the throat and tongue so the large whale dies from loss of blood and it is then eaten. But there is no documented evidence of an unprovoked attack by a killer whale on man.

They are found world-wide, mostly in coastal and cooler regions (Map 62). They usually live in small groups or 'pods' of less than ten, but larger numbers have been seen.

Pseudorca crassidens — the false killer whale

This whale is long and slender, growing to about 5.5 metres (18 feet). The head is small and rounded. The body is black except for a lighter patch on the belly; the flippers have a hump on the forward edge, and there is a small, curved fin. They are found world-wide, usually offshore in large schools, in tropical, sub-tropical and warm temperate waters (Map 63). False killer whales will come towards a ship and are the largest of the 'blackfish' to bowride.

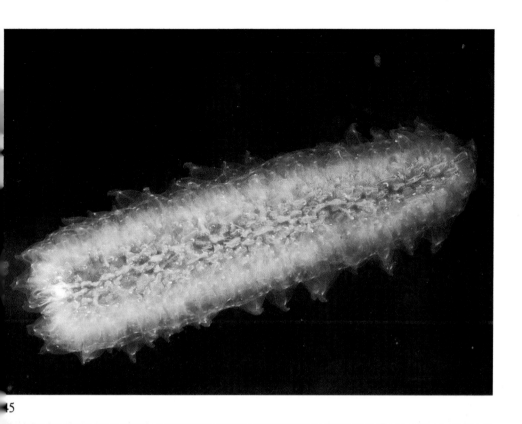

45

6

47

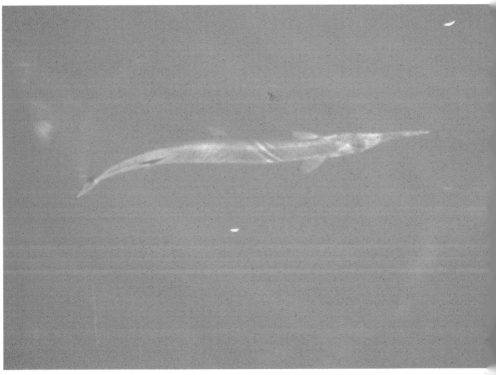

48

49

50

51

52

53 a

3 b

54

55

56

57

58

59

60

61

62

63

64

Pygmy killer whales grow to less than two and three-quarter metres (nine feet). The head is rounded and the slim elongated body is very dark grey or black, with lighter grey along the flanks. White 'lips' and a white lower jaw are often seen. The curved fin is up to 38 centimetres (15 inches) tall. They are found in tropical and warm temperate waters (Map 64).

Feresa attenuata — the pygmy killer whale

Male pilot whales grow to over 6 metres (20 feet); the females to not more than 5.5 metres (18 feet). The head is large and rounded, and the body is black with a heart-shaped greyish throat-patch drawn into a line running backwards. The flippers are long — about one-fifth of the body-length — and the tail is flattened to produce a keel on the top and bottom. The fin is one of the main features. It is broad, thickset and curved.

Longfin pilot whales are found throughout the North Atlantic from North Carolina to the Mediterranean, Greenland and Norway, and in the cool temperate waters of the southern ocean (Map 65).

Pilot whales are the commonest whales to strand on the shore, sometimes in large numbers.

Globicephala melaena — the longfin pilot whale (blockfish)

These males grow to about 5.5 metres (18 feet); females to about 4 metres (13 feet). They are difficult to distinguish from the longfins. They are black, with the thickset fin and grey throat patch, but the flippers are shorter and boomerang-shaped, and there is sometimes a greyish 'saddle' mark behind the fin. Shortfin pilot whales are found world-wide in tropical and warm waters, but there is a slight overlap with the longfins (Map 66). Sometimes the North Pacific shortfin pilot whale is called *Globicephala sieboldi*.

Globicephala macrorhynchus — the shortfin pilot whale (blockfish)

Melonheaded whales grow to three metres (ten feet) long, and have a head shaped like a rugby ball. The body is elongated and narrow, with a black or dark grey back and a lighter belly. The throat has a light grey anchor patch and the curved fin is in the middle. These whales are found in the tropical oceans, generally in schools of 100 to 500 (Map 67) and although they are not known to bowride they may come towards a ship to 'investigate'.

Peponocephala electra — the melonheaded whale, electra dolphin or many-toothed blackfish

There are about 22 dolphin species found all over the world and even in some rivers. Generally they are between two and four metres (six to twelve feet) long with an obvious snout and a curved fin. Most have a very sensitive echo-location system for communication, and for catching food which is mainly fish and squid. Many dolphins are very active and will approach a ship and bowride.

Family 6: Delphinidae

The bottlenose dolphin grows to four metres (13 feet) long. Generally the back is dark, the flanks lighter, and the belly creamy but there is much variation (Colour Plate 55). Those from the Florida area are lighter grey; those from the north-eastern Atlantic and Mediterranean are a brownish grey; while the tropical east Pacific species are dark to almost black. The

Tursiops truncatus — the bottlenose dolphin

belly often has a pink tinge.

There are two types: one large and robust with a short snout found in cooler waters; the other is slender, with a longer snout and is found living in warm waters (often called *Tursiops aduncus*) The fin is tall and, like the flippers, is curved and pointed Bottlenose dolphins are distributed world-wide, mainly living within the continental shelf, over shallow banks and offshor islands (Map 68).

Delphinus delphis —
the common dolphin
or saddle-back (USA)

Common dolphins grow to two and a half metres (eight feet) They show a large variation in colour, but the main feature is a horizontal figure-of-eight flank pattern (Colour Plate 24). The back is dark, often brownish grey. and the belly is white. The forward part of the figure-of-eight may be from a pale lemon to orange colour; the after part is usually greyish. A distinct dark strip extends from the flipper to the front of the lower jaw and from the eye to the lower jaw. The flippers and flukes are dark and the curved fin may have a lighter area in the centre. The snout varies in length and there may be a long snouted variation in the Arabian Sea.

Common dolphins are found world-wide in tropical and warm temperate waters including the Black Sea (Map 69).

Stenella spp.

There are three types of dolphin species in this group (1) spinner dolphins with a long slender snout; (2) spotted an bridled dolphins with a shorter snout and spots; (3) stripe dolphins with a snout size between the other two and a dar Plimsoll line along the flanks. However, the classification of thi group is difficult, and much more detailed information i required.

Stenella longirostris —
the spinner dolphin

Spinner dolphins grow to two metres (six feet). There is a larg amount of variation, but four types have been distinguished: Costa Rican and eastern form found off central America, whitebelly form found in the mid-oceans and a Hawaiian forn in the western Pacific. But much more detailed observation o the various types is required. Generally spinners have a larg number of teeth and a long thin snout. The flippers and fins ar large and pointed. In the Costa Rican dolphin the fin i relatively straight but it is curved in the other spinners.

Spinners are active dolphins. They get their name from thei habit of leaping high, then twisting and turning in mid-air befor re-entering the water head-first, tail-first or even horizontall creating a loud splash. They are found world-wide in th tropical oceans (Map 70).

Stenella attentuata —
the tropical, spotted
or bridled dolphin

This dolphin grows to two metres (six feet). Different group have been described including coastal, offshore and Hawaiia variations. They travel in large schools, are fairly robust, an have a moderately short snout. They are found world-wide i the tropical areas (Map 71).

The Atlantic spotted dolphin grows to two and a half metres (eight feet), and may be mistaken for a bottlenose, especially in the immature stage when they have not developed their spots. The spotted dolphin has a longer snout. The fin is curved and the body is dark purplish-grey on the back with lighter grey sides and belly. The mature animal has many white spots — the older the animal, the more the spots — and the line of the lips is white. They are found throughout the tropical and warm temperate waters of the Atlantic, Caribbean and the Gulf of Mexico (Map 72).

Stenella plagiodon — the Atlantic spotted dolphin

Striped dolphins grow to about two and a half metres (eight feet), and have a slim body with a distinctive colour pattern. The back is dark blue-grey, sometimes a brownish colour, and the sides are a lighter colour. Beneath this lighter region a thin black line extends from the eye to the genital region. Below this line the belly is white. A dark line also runs forward from the flippers to the eye. The snout is moderately long. These dolphins are found world-wide in temperate and tropical waters (Map 73).

Stenella coeruleoalba — the striped or euphrosyne dolphin

Little is known about this species — if indeed it is a separate species. It grows to about two metres (six feet) long and its distinctive feature is a stripe running forward from the blowhole to the snout and continuing down the snout. The body has a dark back with a grey figure-of-eight pattern along the flanks, and a pinkish belly. This is a rare species and it will be worth taking special note of any bowriding dolphins in the warm waters of the Atlantic where this animal may occur (Map 74).

Stenella clymene — the clymene dolphin

This group of dolphins all have a flat short snout. There are six species, three in the northern hemisphere and three in the south.

Lagenorhynchus spp.

The whitebeak grows to over three metres (ten feet). It has a white snout seven to eight centimetres long (about three inches) and a deep robust body. The back is dark grey with a greyish white area in front of the flippers that sometimes extends over the blowhole. Another greyish patch extends from the creamy belly up behind the fin and across the back. The fin is tall — more than 33 centimetres (13 inches) — and curved.

Lagenorhynchus albirostris — the whitebeak dolphin

These dolphins are found in the North Atlantic from Cape Cod to Greenland and the United Kingdom to the Barents Sea (Map 75). They are found in the North Sea in winter as well as off the Cornish coast and western isles. Pairing takes place in the southern part of their range and most calves are born in the northern feeding grounds in midsummer after a gestation period of ten months.

This dolphin grows to about two and three-quarter metres long (eight feet) and has a distinctive shape and colour. The snout is short, and the body elongated and flattened giving the impression of keels on the top and bottom aft of the fin. The back is a dark grey to black and the belly is white. The sides are grey but there is an elongated white lens-shaped patch below and behind the sharply curved fin. A khaki coloured patch

Lagenorhynchus acutus — the Atlantic whitesided dolphin

extends along the sides from behind the fin to the flukes. The flippers are dark grey and a grey line runs from the flipper to the jaw. The eye is dark and prominent and the snout is short with the upper jaw black, the lower jaw lighter. These identification points can all be seen when the animal rolls. Like the whitebeak, this dolphin is found in the North Atlantic although it does not range as far north (Map 76).

Lagenorhynchus obliquidens — the Pacific whitesided dolphin

These sleek, powerful dolphins grow to about two and a quarter metres (seven feet). Almost no snout is visible because they have a gently sloping 'forehead'. The back is dark, the belly and throat white, and two elongated grey-white patches are visible on the sides (Colour Plate 56). The flippers and flukes are dark grey and the fin is curved and dark at the front but grey behind. These dolphins are spectacular jumpers, making them popular in United States marine parks. They are found in the North Pacific from the Aleutian Islands to Baja California in the east and Kamchatka and the Kuril Islands in the west (Map 77).

Lagenorhynchus obscurus — the dusky dolphin

Dusky dolphins grow to about two metres (six feet) and are similar to the Pacific whitesided dolphin, with the same short snout. The topside is black and the belly white but the flanks are grey with two parallel bars of black sweeping down to the rear from under the fin. The flippers, flukes and tail are also dark. A grey line runs from the flipper to the mouth and the eye is black. The fin, in the centre of the back, is dark with a grey colour to the rear.

These dolphins are mostly coastal but they are occasionally seen offshore in the cool temperate waters of the south including South Africa, southern Australia, New Zealand and the southern part of South America (Map 78).

Lagenorhynchus australis — the Peale's dolphin or black-chinned dolphin

This southern dolphin grows to just over two metres (six feet), has a short snout, a sloping front and is dark grey on the head, back and tail. There is a grey area forward and above the dark pointed flippers, and a smaller grey area towards the tail. The throat and belly are white, and the fin is tall and sharply curved. Peale's dolphins are found only around Chile north to Valparaiso, southern Patagonia and the Falkland Islands (Map 79).

Lagenorhynchus cruciger — the cruciger or hourglass dolphin

This dolphin is a similar shape to other southern species, but the colour pattern is a dramatic black and white in the shape of a crucifix when seen from above. The back is dark from the snout to the flukes and a line runs from the eye to the flippers. Another black band sweeps up to the fin from the flippers and down to the underside. Above and below these bands the body is white which makes an hourglass shape on the sides. The tail fin is black, slightly pointed and curved as are the flippers. Hourglass dolphins are found all around the southern ocean usually far offshore in the cold waters north and south of the Antarctic convergence (Map 80).

Cephalorhynchus spp.

All the four species of dolphins in this group are found in th

outhern hemisphere. They are small — about one and a half
netres (five feet) long.

his stocky dolphin has a small triangular head with a step at the
:art of the snout. The flippers are small, narrow and rounded, as
 the stumpy fin (Colour Plate 57). The head, upper and lower
ıws, shoulder, flippers and fin are black; the throat is white,
nd a white diagonal area extends across the back in front of the
·n and down the sides, dividing into two bands under the belly.
'he rear is black.
   This dolphin is found in the coastal waters, creeks and
hannels of the southern tip of South America. It is very
ɔmmon around the Falkland Islands and is said to occur around
ɔuth Georgia. It is recorded off Kerguelen Island and may be
ff other sub-antarctic islands (Map 81).

Cephalorhynchus commersoni — the Commerson's or piebald dolphin

his little-known dolphin is all black except for a few white
ıtches on the throat and belly. It has no snout and the fin and
ippers are small. Black dolphins are rare and are found around
ɔuthern Chile, in coastal waters, from about 37° south to Tierra
·l Fuego (Map 82).

Cephalorhynchus eutropia — the Chilean or black dolphin

his dolphin also has no discernible snout, narrow flippers and a
nall triangular fin. It is mostly black but with a white belly; the
hite extending upwards on the flanks in front and behind the
ippers. They are rare and are found in the coastal waters
 southwestern Africa from about 22° south to Capetown
Лap 83).

Cephalorhynchus heavisidi — the Heaviside's dolphin

his species is basically greyish, but the head, front of the lower
·w, fin and flukes are darker than the flanks. The throat, belly,
·d under the flippers is white with a white flash extending up
·d back along the flank. The fin is small and rounded. The
ıimals will often come close to small boats and will allow
·emselves to be approached by bathers. In front of a ship they
·im more like seals rather than leaping out of the water. They
·e found in the coastal waters of New Zealand, especially to
·e east of South Island (Map 84).

Cephalorhynchus hectori — the Hector's or New Zealand dolphin

ıis species looks very much like its relative, the southern right
hale dolphin (*see below*). The body is elongated and very
·nder, growing to two and three-quarter metres (nine feet).
ıe snout is not pronounced and there is a faintly discernible
·ck. The animal is black except for a white tip to the lower
·v, white extremities to the undersides of the flukes, and a
ıite crest patch between the flippers that extends in a broad
·e down the underside. There is no fin and the flippers are
·rrow and curved. It is found in the North Pacific (Map 85).

Lissodelphis borealis — the northern right whale dolphin

ıis is very similar to the northern species but only grows to
·o metres (six feet). It also has no fin. The back and top of the
·les are black; the belly, lower sides and head are white
 olour Plate 58). They are found all around the southern
·eans, but mainly along the Antarctic convergence (Map 86).

Lissodelphis peroni — the southern right whale dolphin

Grampus griseus —
the Risso's dolphin,
grey grampus

This large and robust dolphin grows to 4 metres (13 feet) an has a small, blunt head with no snout. Over the front of th head is a central crease. The flippers are long and curved, an the fin is large — 40 centimetres (16 inches) — and also curve and pointed.

The colour depends on age and area, with variations of dar grey to almost black, and light grey to almost white. The body : marked with a light dappled effect and is often heavily scarre The flippers and flukes are darker than the rest of the body There is a light heart-shaped patch under the throat whic continues in a line down the chest and belly. The eyes ar noticeably dark. This species should not be confused with an ol name for the killer whale which was the grampus. Risso dolphins are found world-wide in temperate and tropical ocea (Map 87).

This animal was the first individual dolphin to be protected b parliamentary legislation. In 1904 the Governor of New Zealan signed an Order in Council 'to prohibit the taking of the fish c mammal known as Risso's dolphin in the waters of Cook Stra or bays, sounds and estuaries adjacent thereto'. This order was t protect a dolphin which, since 1888, had been regularly meetir ships, escorting them and bowriding in a stretch of Cook Stra opposite Pelorus Sound. 'Pelorus Jack', as the animal came to k known, had this order renewed twice and became a great touri attraction. He, if it was a male, was last seen in April 1912.

Orcella brevirostris —
the Irrawaddy dolphin

This dolphin is named after a river in Burma where it has bee found 900 miles from the sea. It grows to two metres (six fee and has an oval head with no beak but a flexible neck. Tl flippers are oval and paddle-shaped, and the fin is small ai triangular. It is a dull muddy grey on the back and sides b lighter below. The dolphin is found in many coastal are estuaries and large rivers around the Bay of Bengal, the Mal: Peninsula and northern Australia (Map 88).

Lagenodelphis hosei —
the Fraser's dolphin
or Sarawak dolphin

This species is unusual in being intermediate between the longe snouted common dolphin, *Delphinus*, and the short-snout *Lagenorhynchus* dolphins. It grows to two and a half metres (eig feet) and is a robust animal with a dark grey back, light gr flanks, and a creamy white lower jaw, throat and belly. There a broad stripe along the sides. It may be confused with tl striped dolphin (*Stenella coeruleoalba*) but this species has a mu shorter snout. The fin is small and moderately curved. Tl flippers and flukes are also small. They are found in mc regions of the Atlantic, Pacific and Indian oceans and may l relatively common in deep waters, but little is known (Map 89)

Family 7: Stenidae

This family contains four species and is often included in tl larger dolphin group.

Steno bradensis —
the rough-toothed
dolphin

This dolphin, which may be up to two and three-quarter metr long (nine feet), gets its name from its furrowed teeth. The he is cone-shaped and the snout is not obvious. Its colour generally dark purplish on the back with light grey sides and white or pinkish-white belly. Pinkish-white blotches may also be seen on the sides, and older animals often have scars. The slightly curved fin is central and the flippers and flukes are dark. Rough-toothed dolphins are believed to live in most tropical warm temperate oceans (Map 90). It is usually seen offshore and has been recorded from the Hawaiian Islands, the eastern tropical Pacific, the Indian ocean and the Caribbean.

Sousa chinensis —
the Indo-Pacific or
Chinese white dolphin

This species grows to about three metres (ten feet) and the body colour is variable. Animals from East Africa, India and northern Australia are dull lead grey on the back and sides but white on the belly. Those from Borneo and the China Sea are almost cream in colour with pinkish brown tips to the fins and flukes. The young are usually dark grey.

These dolphins have a long slender snout. The fin is on a hump which may be prominent in the East African and China Sea animals but less so in the Australian ones. They are found in the coastal waters of South and East Africa, the Red Sea, the Persian Gulf, Indonesia, North Queensland, Taiwan and the South China Sea (Map 91).

Sousa teuszii —
the Atlantic humpback
dolphin

Atlantic humpbacks grow to just over two metres (six feet). They have a long snout, and a grey body with a lighter belly. The fin is rounded and located above a distinctive hump which is not so prominent in young animals. This dolphin lives in the coastal waters of tropical West Africa from Cameroon to Mauritania (Map 92). It will enter harbours and large river estuaries and may mix with bottlenose dolphins at sea.

Sotalia fluviatilis —
the tucuxi or estuarine
dolphin

This is one of the smallest dolphins, growing to one and three-quarter metres long (five feet). Its body-shape is similar to that of the bottlenose. Its colour varies from the grey of the lower Amazon River form to the brownish grey of the estuarine and coastal animal. Both are pale underneath. The snout is long and the fin and flippers are curved slightly.

The tucuxi lives mainly in the Amazon and its tributaries, but also on the coast and in estuaries along northwestern South America from Colombia down to about 24° south (Map 93).

Family 8:
Phocoenidae

These small, squat animals — usually with a small triangular fin and short snout — are the porpoises. The separate species can be distinguished by their different geographical areas.

Phocoena phocoena —
the common or
harbour porpoise

Common porpoises grow to about one and three-quarter metres (five feet) long, females being slightly larger than males. They are small and plump with narrow flippers and a small triangular, central fin. The back, tail, flippers, flukes and lower lip are all black, but the lower jaw, throat, chest and belly are white (Colour Plate 25). A grey patch extends above the flipper from the eye to just under the fin.

They are coastal animals, found in the North Atlantic from West Greenland and the Davis Strait to Cape Hatteras in the west, and from the Barents Sea to Senegal in the east (Map 94). In the Pacific they are found on the North American coast to

Alaska and Baja California, and on the Asiatic coast to Japan. An isolated population in the Black Sea is described as a subspecies, *Phocoena phocoena relicta*, but the animal is no longer found in the Mediterranean.

Phocoenoides dalli — the Dall's porpoise

This porpoise grows to just over two metres (six feet) and is also rather plump. The head and flippers are small and the fin is triangular and slightly forward of the mid-point. There are two variations of colour. In one, the head and body are all black except for the white trailing edge of the fin, flukes and a large patch on the flanks: the belly is also white. These are found in the western Pacific around Japan. The other type has a much larger white patch which extends forward to include the chest. This is found in the eastern Pacific along the Canadian and United States coast. Totally black and grey/brown variations have also been described. Dall's porpoises are found in an arc from Japan to Baja California (Map 95).

This porpoise is easily recognisable as not only will it readily come towards a ship but, when swimming along — porpoising — it creates a characteristic V-shape spray.

Phocoena sinus — the cochito or Gulf porpoise

This is similar to the common porpoise but only about one and a half metres long (five feet) and brownish with a small curved fin. The cochito lives only in the upper Gulf of California (Map 96).

Phocoena dioptrica — the spectacled porpoise

This porpoise is elongated and grows to over two metres (six feet). It is dark grey on the back and flukes, has a dark ring around the eyes, and a dark lip-line. It has white upper and lower jaws and white on the lower half of the body including the flippers. A grey line runs from the flippers to the mouth. This species lives around the tip of South America, Tierra de Fuego, the Falkland Islands and South Georgia (Map 97).

Phocoena spinipinnis — the Burmeister's or black porpoise

This is similar to the common porpoise, growing to one and three-quarter metres (five feet) long, but it has a slightly more pointed head. It is dark grey to black, with lighter grey on the belly. The flippers are small and dark; the fin is wedge-shaped. These porpoises are coastal and are found around South America from 18° South to 40° South on the Pacific side, and from Uruguay to Patagonia on the eastern side (Map 98).

Neophocaena phocaenoides — the finless porpoise

There are two subspecies of finless porpoise; *Neophocaena phocaenoides phocaenoides*, from the Indian ocean, and *Neophocaena phocaenoides asiaeorientalis*, from China and the Far East. They both grow to about one and three-quarter metres (five feet) and have a blunt rounded head without a snout. Far Eastern animals have a more pointed head when seen from above. They have no fin but a dark ridge of tiny raised pimples runs along the back from midway between the flippers to the tail. The body is light grey shading to white on the belly.

Finless porpoises live in coastal regions and in creeks and major rivers of Pakistan, India, Indonesia, China, Korea and Japan (Map 99).

This family includes the beaked whales. They are all medium-sized whales with a prominent snout and a pair of grooves running down the throat. The blowhole is crescent-shaped with the 'arms' pointing forward, and the fin is usually well back on the body. Their main feature is that, except for the Tasman whale, they all have four teeth or less.

Squid form the major part of the diet though they will also eat fish. Perhaps the best way to identify these species is by their geographical area, but unfortunately this is not simple. Very little is known about the animals and some are only known from skulls. Many more observations are required.

This species grows to about 10 metres (32 feet). The 'forehead' is prominent and the body is well rounded, tapering towards the flukes, which have no central notch. The body is slate colour or brownish-grey, shading to lighter below, and usually covered with scars. The snout has two large teeth on each side of the tip of the lower jaw. The fin is small, curved and placed well back. Distribution is vague but the species is known from New Zealand, Australia, South Africa and the Falkland Islands (Map 100).

The body shape and colour are similar to the southern species but this animal grows to almost 13 metres (43 feet); they also have scar marks. It is found only in the Bering Sea, off Japan, and in the waters of central California where it is relatively common between June and October (Map 101).

Northern bottlenose whales grow to 10 metres (32 feet), have a robust body, a bulging 'forehead', and dolphin-like snout. The animal is usually dark brown with a lighter brown belly, although some individuals are shades of grey. The fin is about 30 centimetres (12 inches) tall. They are found from the north-eastern United States and north-western Europe to the Arctic, normally in deep water outside the continental shelf (Map 102).

This whale is very similar to the northern species but only grows to 8 metres (26 feet). It occurs around the entire southern ocean (Map 103).

This whale is similar in shape to other whales in this family but it has more teeth and only grows to 6 metres (20 feet). It is dark grey with scar marks and a lighter belly. The animal is very rare and known only from a few stranded specimens, but it is believed to live around the Antarctic convergence (Map 104).

Very little is known about this species as it has never been seen alive: only two stranded specimens have been recorded. It is about 7 metres long (23 feet) and has two teeth in the lower jaw (Map 105).

Family 9: Ziphiidae

Berardius arnuxi — the southern giant bottlenose whale, Arnoux's whale or southern four tooth whale

Berardius bairdi — the North Pacific giant bottlenose whale, Baird's beaked whale or northern four tooth whale

Hyperodon ampullatus — the northern bottlenose whale

Hyperodon planifrons — the southern bottlenose whale

Tasmacetus shepherdi — the Shepherd's beaked whale or Tasman whale

Indopacetus pacificus — the Longman's beaked or Indo-Pacific beaked whale

Ziphius cavirostris — the Cuvier's or goosebeak whale	These whales are about six metres long (20 feet) — females being larger than males. The body is robust with a small head and not such a prominent snout. The Atlantic animal is a slate grey with a lighter belly, but in the Indo-Pacific it is a dark rust brown. Both types have obvious white- or cream-coloured spots, and scars, and older males often have a cream-coloured head. The fin is about 38 centimetres (15 inches) tall and curved. They are found world-wide in all temperate and tropical seas (Map 106).
Mesoplodon spp.	They are similar in appearance, between 4 and 7 metres long (13 to 23 feet), and are rarely seen. In fact they are mostly known from strandings, but it is possible that they are more common but found in deep waters and shy of shipping. They are difficult to identify at sea and many more careful observations are required.
Mesoplodon hectori — the Hector's or skew-beaked whale	This is known only from a few skulls found in Tasmania, New Zealand, the Falkland Islands and South Africa (Map 107).
Mesoplodon bowdoini — the deep crest beaked whale or splay tooth beaked whale	Also only known from New Zealand, Tasmania, Australia and the Kerguelen Islands (Map 108).
Mesoplodon layardi — the straptooth whale	In addition to Australia, New Zealand and the Falklands, this species is also known from South Africa (Map 109).
Mesoplodon grayi — the scamperdown whale	Found from South Africa, South Australia, New Zealand, Chatham Island and Argentina (Map 110).
Mesoplodon carlhubbsi — the Hubb's beaked whale	This species is found in the temperate waters of the North Pacific (Map 111).
Mesoplodon stejnegeri — the sabretoothed whale or Bering Sea beaked whale	As its name suggests, this whale is found in the subarctic waters of the North Pacific (Map 112).
Mesoplodon ginkgodens — the ginko-tooth whale or Japanese beaked whale	This species is found in the Pacific from California to Japan and Taiwan (Map 113), and around Sri Lanka.
Mesoplodon mirus — the True's beaked whale or wonderful beaked whale	There may be an isolated population of this species off South Africa, but the majority occur in the temperate waters of the North Atlantic, from Florida to the United Kingdom (Map 114).
Mesoplodon europaeus — Antillean or gulfstream beaked whale	This whale is found in the western Atlantic, off the eastern United States and in the Caribbean; there is one record from the English Channel (Map 115).

The densebeak whale is found in tropical and temperate waters of all oceans (Map 116).

Mesoplodon densirostris — the densebeak whale

This species is found in the cool temperate waters of the North Atlantic from Nova Scotia to southern Norway and the Bay of Biscay (Map 117).

Mesoplodon bidens — the North Sea beaked whale or Sowerby's whale

Both species in this family are found in the Arctic, and they are easily distinguishable.

Family 10: Monodontidae

Male narwhals grow to about five metres (16 feet), the females to about four metres (13 feet). The body has a small rounded head with no snout. Males have a prominent tusk, actually one of the front teeth, up to two metres (six feet) long growing out of the front left upper lip. The tusk is possibly used for courtship displays and for digging in the sea-bed in search of crabs and shrimps. The narwhal also eats squid and fish. Females have no tusk. The plump body has no fin, but has a low undulating ridge along the back. The flippers are small and rounded. They are mottled dark grey/brown on the back and sides; the head and ends of the flippers and flukes are dark grey. The belly is white. They are found around the Arctic (Map 118).

Monodon monoceros — the narwhal or unicorn whale

Belugas grow to 5.5 metres (18 feet) long. The body is cylindrical and fat and has no fin but a ridge along the mid-back; the head is rounded and has no prominent snout — the 'forehead' tends to bulge in males. There is an obvious neck. The flippers are small, rounded and, in the male, upturned.
Adult whales are pure white except for the top of the back which may be greyish. The tips of the flippers and flukes may also be grey. At birth the calves are a dark grey/brown which fades quickly after the first year, gradually becoming bluish grey by approximately the sixth year. The white colour after this age usually signifies sexual maturity. Belugas are found in the Arctic and subarctic. They are common in the Saint Lawrence Seaway and Saguenay rivers at certain times of the year (Map 119). In 1966 one swam up the River Rhine in Germany and they are known to travel 700 miles up the Yukon in the summer.
This is not the white whale, Moby Dick, that was hunted by Captain Ahab in the story by Herman Melville. It is more likely that this was a sperm whale.

Delphinapterus leucas — the beluga or white whale

This family contains the four species of freshwater dolphin which may be the most primitive of the cetaceans. They can be identified according to the rivers in which they live. They are all small, with long slender snouts and bulging 'foreheads'. They have a noticeable neck and can turn their heads sideways, up and down. Because they live in murky river waters their eyes have been much reduced and the Ganges and Indus river dolphins are actually blind. They compensate by having a very elaborate sound system to communicate and find food. They mostly feed on catfish.

Family 11: Platanistidae

Platanista gangetica —
the Ganges and Indus
River dolphins or susa

These dolphins are found in the Ganges, Brahmaputra and Indu
River systems (Map 120). They grow to about two to two and a
half metres (six to eight feet) long. The name 'susa' is an Urdu
word that refers to the sighing noise made by the dolphin when
breathing. It is possible that there are two subspecies; *Platanist
gangetica gangetica* living in the Ganges and Brahmaputra Rivers
and *Platanista gangetica minor* living in the Indus River. Thi
dolphin's survival is threatened by the construction of dams and
barrages for irrigation systems. The animal is noted for being
constantly on the move as it has not been seen to rest either on
the bottom or at the surface. They are continuously emitting
sounds, although with frequent interruptions.

Lipotes vexillifer —
the Yangtze river
dolphin or baiji (China)

This species is found only in small groups or singly in the lowe
reaches of the Changjiang (Yangtze) River (Map 121). It is whit
or light grey; it has no predators and is shy, but some anima
are killed by boats and general river traffic. It is revered an
protected rather like the giant panda, and although number
must be low, a few are kept under control conditions in
research pool at the Academica Sinica in Wuhan where it
hoped they will breed.

Inia geoffrensis —
the Amazon River
dolphin or bouto

There may be two subspecies; one from the River Be
in Bolivia, the other from the Amazon and Orinoco Rive
(Map 122). The snout of this dolphin has a few stiff hairs.
problem is posed by people settling in the area and huntin
them. Prior to this, the locals left them alone because of a beli
that they changed into beautiful women at night and entice
men into the water to drown. But an increasing threat today
posed by capturing them for export, during which many die.

Pontoporia blainvillei —
the La Plata River
dolphin or franciscana

This species is unusual in that it also lives in the estuary of th
River Plate and on the coast, but all observations have bee
within 16 nautical miles of the coast. It is also found in the R
Paraguay and Rio Uruguay (Map 123). It is about one and a ha
metres long (four and a half feet) and has the characteristic lor
snout and many teeth. It is a soft brown buff colour with a pal
mustard-yellow belly.

THE INTELLIGENCE OF CETACEANS

The intelligence of these animals has long been the subject
much debate, and there are many anecdotal stories of the
behaviour in the wild. Many ways have been tried to measu
intelligence, but it is difficult to quantify and define — even
humans. Brains have been weighed, measured and compared
try to get some idea of relative intelligence. But comparing bra
size does not provide the answer because of large differences
body size. Even comparing the brain weight to body weight ra
does not give a good measure. Nor does measuring the degr
of folding, which reflects the capacity of the brain.

Strict comparisons do not really tell us much about cetacea
intelligence. But the fact is that there are large differences in t
demands made on the brain and consequently in the developme

of its different areas. These animals live in a totally different environment — one that demands different abilities. In the sea, sound and touch are important and the areas of the brain dealing with these senses are very well developed. On land, sight and smell are also important, and the human brain reflects this requirement.

But it is perhaps the social ability that is most highly developed, especially in the toothed whales. They have complicated 'family' relationships, and peer groups within the schools, with dominant adults and young immatures. They will rescue injured companions, assist at birth, and even help people. There are also times during the day when dolphins seem to take time off just to 'play'. Sound is important in these activities and conveys the particular 'mood' of an individual. Certainly if they have the ability to kill fish by sound (see page 101) then it would be important to keep this under control.

The ability to communicate has also been used as a measure of intelligence, and people seem preoccupied with trying to speak to dolphins. But while dolphins can perform tricks to spoken commands, and have the ability to mimic sounds, it is impossible to speak to a dolphin in a common language or translation with any real exchange of information.

Whales and dolphins are very well adapted to their environment. But, despite their obvious abilities, many species have not learnt to avoid trouble. For instance, there are frequent reports of whale collisions (see below). In some cases an animal has deliberately and fatally attacked a ship. In addition, whales that were hunted did not learn to avoid ships, and some continued to approach vessels — to their misfortune. Seen in human terms, this is not very clever.

The intelligence debate always arises when discussing strandings. This is where, for no apparent reason, an individual or group will beach themselves and appear to commit suicide. There have been many suggestions to explain this behaviour. One theory suggested that the animals were dying and wanted to commit suicide; another said that the whale was dying and wanted to return to the land from whence its ancestors came. Such stories have been discounted.

Strandings may be caused by errors in the whale's navigation or echo-location system. There are some particular places in the world where whales have been known to strand over many centuries, but these may just be difficult areas for them to navigate close to the shore. But in some cases it has been found that stranded animals have been infected with a parasite in the inner ear. In addition, strandings sometimes take place where there is a gently sloping soft sea floor, which would not give a clear echo. Whether it is an infection or an unclear echo, the animal finds itself in difficulties and 'screams for help'. Its companions come to help and they, too, get into trouble, and so it continues until the whole group, sometimes consisting of many animals, becomes stranded. This also explains why, once an individual is towed away, it will immediately return to help its companions. But this is not the full story, and there are complications. Strandings are, in fact, rare events; and there are different kinds of strandings. Sometimes killer whales may be stranded after chasing seals or penguins, or schools may be seized by blind panic when being attacked, or when finding themselves in shallow water. It is possible that the weather, or the character of the water, may have altered and that this may result in disorientation. Whatever the cause, the sight of helpless animals stranded must be among nature's most tragic.

If a stranded animal is found, it is important to shield it from the sun and keep its skin wet and cool — especially the flippers, which can absorb the heat quickly. Then, if possible, return the animal to the sea after taking full measurements and a photograph or drawing. For a larger species perhaps the only way is to refloat it on a high tide.

WHALE COLLISIONS

Collisions between ships and whales have been reported. There may be several reasons. Firstly, the animal may have been asleep. Some of the larger whales sleep in the warmer waters, floating about two metres below the surface, but surfacing gently to breathe. The sleep is not deep in human terms. Dolphins have been seen to have quiet periods but usually waken at the slightest noise. Secondly, an animal may be sick or diseased and unable to take avoiding action. Finally, it is possible that a whale may be dead on the surface before the collision.

A HISTORY OF WHALING

Whale hunting has been carried out by the Eskimos since 1500 BC, and by the Norwegians since AD 890. These early whalers obtained food, clothing, weapons and lighting from the whales. They hunted the Greenland right whale because it is a slow swimmer and floats when killed. Early whaling was shore-based but, with the development of furnaces and boilers on ships, whaling went to sea. Many other nations, including the English, Dutch and Americans began hunting the right, grey and sperm whales extensively.

In time two developments greatly improved whale hunting. The first was the invention of steam power. This enabled bigger and faster ships to hunt the larger faster whales — rorquals — and to penetrate the pack ice to reach more whaling grounds.

The second development was the explosive harpoon, weighing 160 pounds, about two metres (six feet) long, with an explosive grenade in the tip. On hitting the whale, the grenade explodes and the animal swims off pulling the connecting rope and the ship gives chase, meanwhile the harpooneer reloads his gun to stop the whale. Using this method it takes between a few minutes and two hours for the whale to die. Other methods such as poisoning or electrocution have been tried, with no success, and the harpoon remains in use today.

Early whalers only wanted the blubber for oil but, as the industry progressed, other products could be obtained: fertilizers, medicines, tennis rackets and umbrellas, baleen for corsets and food for pets and humans. Ambergris is a substance produced by the sperm whale during digestion. It is a brown, waxy aromatic substance with squid beaks embedded in it, and was used as a fixative for perfumes and make-up.

There was much money to be made, and between 1925 and 1930 the industry developed the factory ship system. The catcher vessels no longer towed their catch to the shore, but could leave them at the factory ship. Spotter planes were used to find the whales.

The factory ship system was very efficient at killing whales: in the 1930 Antarctic season (December to April) 41 factory ships killed over 37,000 whales — nearly 28,000 of them blue whales. In 60 years about 330,000 blue whales were killed and it was finally made a protected species in 1966. But by this time the whaling industry was supported by other species because blues were too difficult to find. Subsequently many other species had to be protected: the southern right, grey, fin, humpback and sperm whales. Greenland right whales may only be hunted by local people in the Arctic.

THE INTERNATIONAL WHALING COMMISSION

In 1937 an international whaling agreement was signed by the major whaling nations limiting the number of vessels, the season to three months, and stipulating a minimum size to allow animals to reach maturity. The participants in this agreement were more interested in regulating whale oil prices to their mutual interest than in the conservation of whales.

In 1946 the International Whaling Commission (IWC) was formed by the major whaling nations: non-whaling nations were not involved as it had not occurred to them that they could play a valuable role. In the early stages the members voted entirely on the basis of self-interest to obtain a short-term economic gain, but eventually many countries (those with the worst records include the UK and Australia) began phasing out their whaling interests — which sometimes meant selling their fleets abroad. Only then, when it was inexpensive, did these nations support politically popular conservation measures. The records of these countries for supporting conservation at the times when it could have been effective are appalling.

The IWC itself was ineffective for several reasons:

1. Instead of adopting a quota for each species the Commission adopted the blue whale unit (bwu) where species are rated by their respective oil yields (1 blue = 2 fin = 2½ humpbacks = 6 sei). This, in effect, condemned the blue whales as the quota could be achieved easier and quicker by catching blues.
2. Quotas for individual nations were not adopted, rather there was a total quota for the Southern Ocean. The Americans were mainly responsible for this in the name of 'free enterprise and competition'. This delayed effective conservation measures for at least 15 years.
3. The IWC did not have 'teeth' because any nation entering an objection within a 90-day period was not bound by the Commission. This veto made it impossible to implement serious conservation measures during the 1950s.

These three points made the IWC virtually impotent for ten crucial years. (There is a comparison here with other commissions such as for the regulation of fisheries — but that is another story.)

The IWC has never had the power to enforce decisions. Even if the commissioners were in full agreement, decisions do not become binding on their own governments until local legislation is passed.

A turning point came in the 1962–3 and 1964–5 seasons when the whaling nations could not even catch the quota they were allowed. In the 1970s the hardline nations, Japan, South Korea, the USSR and Canada came under criticism for two main reasons. The first was growing support for a complete moratorium on all whaling. Secondly, an increasing number of non-whaling nations were entering the IWC. This perhaps marks the beginning of an awareness that whales do not belong to any nation but are part of the marine world.

In July 1982 the IWC voted in favour of a moratorium on commercial whaling effective from 1985. In November of the same year Japan, Norway, Peru and the USSR lodged their objections. But the regulations only apply to those nations in the IWC; a significant number of whales are taken by countries who are not members.

Another problem is the incidental catch of whales. This occurs in two ways. Firstly, some animals are taken deliberately but 'in passing'; for instance, some fishing expeditions hunt the occasional whale or dolphin while fishing. This is relatively easy to legislate and control by the imposition of fines or confiscating gear. But the slaughter of thousands of bottlenose dolphins and flase killer whales that are herded ashore and clubbed to death by the fishermen of Iki Island in Japan has recently received attention. The fishermen claim they lose 30 million dollars a year because the animals compete with them in catching cuttlefish and yellowtails. The fishermen have tried driving them away using the sounds of a killer whale or those of a distressed dolphin, but the animals become accustomed to these noises. But what is the full story? Could it be that an increase in fishing intensity by man has forced the dolphins to compete for the remaining fish? In any case, is the maximum profit motive so important that the killing of thousands of dolphins becomes a necessity?

The second type of incidental catch is caused by accident, which is more of a problem to control. Several species are causing concern:

1. The spinner, spotted and common dolphins caught in tuna nets by US fishermen. The dolphins indicate where shoals of tuna are to be found and in the course of purse seine netting between 113,000 and 244,000 were killed each year in the 1960s and 1970s. The fishermen have tried letting the end of the net fall back — backing down — to let the dolphins escape, but very few manage to do so. Panels were then built into the net to enable the dolphins to escape and this has had a moderate success. It is not known whether the populations are in danger but the neeedless slaughter of hundreds of thousands of dolphins every year is a tragedy.

2. Between 10,000 and 20,000 Dall's porpoises are killed each year by the Japanese salmon gill netting operations. The nets are set overnight and the porpoises seem unable to avoid them. Decoy models of killer whales and playbacks of their sounds have been tried but the animals become used to these and ignore them.

3. Harbour porpoises get caught in salmon drift nets used by Greenland and Canadian fishing vessels. It is estimated that between 2,500 and 2,700 are caught each year. This could

have a severe effect because this species is short-lived and does not calve every year.

4. An unknown number of La Plata dolphins are caught in nets set for sharks off the south east coast of South America.

5. Humpback whales are trapped in fishing gear off the coast of Newfoundland. This problem dates from the 1970s when the whales began to come closer to shore, possibly due to the fall in number of offshore fish (capelin). The coastal fisheries use set nets and cod traps, and between 1969 and 1978 at least seven humpbacks and ten minke whales have been killed. This could have a severe effect on the Atlantic humpback population which is very small — between 1,200 and 1,500.

Much more work needs to be done on fishing methods to decrease this incidental catch. Public awareness and opinion to put pressure on authorities to carry out this work is important, otherwise the outlook for some species, particularly the harbour porpoise and humpback whale, is grim.

THE CONSERVATION OF WHALES

Whales reproduce slowly and they have very few natural enemies so they have not evolved to cope with large scale hunting. Hence for some species it will be a long time before we know if we have protected them in time.

Conservation may be seen from different viewpoints. Some people favour conservation so there will be whales for future exploitation. However, today there are synthetic products for everything the whales used to give so to continue whaling is barbaric and unnecessary in our society.

Another point is that whales are part of the marine system — the web of life; removing them will upset this web. The decline of whales has resulted in an increase in the amount of krill. This may be the cause of an increase in some populations of seals and penguins. The krill is being viewed as a prospective fishery. If this develops and the krill are fished intensively this could add another problem to the recovery of whales.

But should not whales be simply allowed to live? It may be difficult to convince the Newfoundland fisherman who has just lost thousands of dollars of fishing equipment, but surely the problems are not insurmountable? It would be a tragedy if the only things we had to show to our children or grandchildren were pictures of whales and tapes of their song.

ORDER 2: THE PINNIPEDIA — THE SEALS

Most people have seen performing 'seals' in zoos and circuses; these are usually Californian sea lions. But seals are often reported from ships, sometimes a long way from land. There are 32 different species, and in many cases they can be identified according to where they are seen; at least it is usually possible to identify the group.

Unlike whales and dolphins, seals must return to land to breed. They haul out onto their preferred beaches and sometimes the adult male or bull gathers a harem of cows. In some species the bull must establish a territory, fighting off challenges from rivals. But before this happens, the females will usually already be on the beaches where they will give birth to pups conceived during the previous breeding season. A short time after the birth — sometimes two weeks — the males mate with the females. But the fertilised egg is held for two or three months before it is implanted in the uterus (womb). This ensures that the pregnancy starts at the right time so that the pups can be born during the correct season and in the right place. Most seals also haul out onto a beach at some time during the year to moult. They feed mainly in the sea on fish and squid.

Seals are divided into two groups. The eared seals consist of the sea lions and fur seals; the walrus is a close relation, although it has no external ear, and it is included in this group. The second group comprises the true seals, and includes northern hemisphere species, the Antarctic, tropical, hooded and elephant seals.

With the exception of the monk seals, these animals are found in areas washed by cool currents; generally they do not live in water warmer than about 20 degrees centigrade. This can be seen if the distribution maps are compared with Figure 1.2. Where they live in the tropics, cold currents penetrate from the polar areas: the Galapagos Islands, for example, are influenced by the cold Humboldt current travelling up the Pacific coast of South America.

1. THE EARED SEALS

Eared seals have large front flippers, without fur, which can be turned outward for support. The back flippers are also large, hairless and can be turned forward underneath the body for movement on land.

Family 12:
Otariidae —
sea lions and fur seals

This family includes five sea lions and eight fur seals. Most species can be distinguished according to their area. Sea lions have a blunt snout and a coat of short coarse hairs covering a small amount of underfur. The fur seals have a more pointed snout and long hairs that cover a very thick underfur which is of considerable commercial value. In many cases this has resulted in a severe depletion of numbers due to hunting.

Eumetopias jubatus —
the northern or
Steller's sea lion

This is one of the largest sea lions, with the males reaching three and a half metres (twelve feet); females grow to two and a half metres (seven and a half feet). The breeding season is between May and June. Steller's sea lions are not often seen in captivity. They are widely distributed in the North Pacific, living mainly on rock islands off the coast, with the centre of abundance in the Aleutian Islands (Map 124).

Zalophus californianus —
the Californian sea lion

These are the best-known sea lions, and the ones usually seen in captivity. They grow to just over two metres (six feet) and are a chocolate-brown colour (Colour Plate 59). They feed mainly on squid, octopus and fish. The performing life of a sea lion may be eight to twelve years.

They occur along the Californian coast, north to Vancouver Island and south as far as Tres Marias Islands, Mexico (Map 125). They are shore-living animals and do not usually wander farther

1an ten miles out to sea. Some of the bulls move north in
vinter, to Vancouver Island. Three subspecies of Californian sea
on have been described, these are: *Zalophus californianus
lifornianus*, which is the Californian animal; *Zalophus californianus
•ollebaeki*, that lives around the Galapagos Islands, and *Zalophus
lifornianus japonicus* that lives in the southern Sea of Japan on
oth sides of the island of Honshu.

outh American sea lions grow to about two and a half metres
:ight feet). They are generally brown but adult males are dark
nd have a paler mane with the belly lightening to a dark
ellow. This sea lion is found on the Atlantic and Pacific coasts
f South America (Map 126).

Otaria flavescens —
the South American
sea lion

ustralian sea lions never move very far from the coast, but they
·e known for their ability to climb inland. They are found
ong the south and southwestern shores of Australia from north
f Perth to Kangaroo Island in South Australia (Map 127).

Neophoca cinerea —
the Australian sea lion

ooker's sea lions prefer sandy beaches, but can be found
nong the tussock grass and bush inland. Males grow to three
etres (ten feet) and are blackish brown; the females are much
;hter and smaller — about two metres (six feet) long. The
·eeding season is from November through to February. They
.t fish, crabs and mussels, and will often regurgitate a ball of
idigested crabs' legs and shells on land. They occasionally eat
·nguins, chasing them along the beach, then taking them out to
a to eat. They are found around New Zealand (Map 128).

Phocarctos hookeri —
the Hooker's sea lion,
or New Zealand
sea lion

1is could be called *the* fur seal because this species has been
·ed most extensively for fur coats. Males grow to about two
:tres (six feet) and are dark brown with a greyish tinge to the
oulders. Females reach one and a half metres (five and a half
·et) and are dark grey on the back, lighter on the belly, with a
·ge of chestnut. Both sexes have a light patch across the chest.
This species lives on the Pacific coast of North America from
·nada to Alaska, along the Aleutian chain, and down the
·stern Pacific to Japan (Map 129). During winter and spring
·ge numbers migrate south, then return to the breeding
·unds in the summer. The limit of the southern migration is
·out 35°N on the Japanese side and at San Diego (33°N) on
· American side. During the migration they are normally about
· miles off land and solitary or in small groups of about ten
·mals.

Callorhinus ursinus —
the northern, Pribilof
or Alaskan fur seal

·e eight species of southern fur seal are all similar. The
·derfur and the short hairs at the base of the flippers are a rich
·stnut colour, and the males have a thick mane of longer
·rd hairs often tipped with white giving them a greyish
·)earance.

·ıth American fur seal males reach about two metres (six feet);
·nales about one and a half metres (five feet). They occur on
Pacific and Atlantic coasts of South America from Rio de

Arctocephalus australis —
the South American
or southern fur seal

Janeiro, around Tierra del Fuego and then north along the coas
of Chile (Map 130). They also live on the Galapagos Islands
There may be three subspecies; the Galapagos form, the Soutl
American form, and the Falkland Island form. Although thi
species lives in similar waters to the southern sea lion, the tw
tend to keep separate — the fur seals preferring rocky shores t
the sea lions' sandy beaches.

Arctocephalus pusillus —
the South African or
Cape fur seal

Male South African fur seals grow to two and a half metre
(eight feet) and females to about two metres (six feet). The
feed mostly on fish, squid and crabs, but rarely dive belov
50 metres (165 feet).

There are two subspecies: *Arctocephalus pusillus pusillus*
the true Cape fur seal: they live along the coast of South Afric
and have been reported swimming up to 700 miles from lan
(Map 131). They wander throughout their range but usuall
return to their birth place for breeding. The second subspecies
Arctocephalus pusillus doriferus, the Tasmanian or Australian fur sea
living along the south-eastern coast of Australia from Eclips
Island in the west to Kangaroo Island in the east (Map 132).

Arctocephalus tropicalis —
the subantarctic or
Amsterdam Island fur
seal

Subantarctic fur seal males grow to two metres (six feet), femal
to about one and a half metres (four and a half feet). The mo
northerly males have a dark grey back, a yellow throat, face an
chest and a brown belly. Females are similar although the yello
is not as bright. Southern females and males are brownish gre
shading to russet under the chin and throat, but the ha
covering the neck and shoulders of the males is long and ragge
looking. They feed on fish, squid and krill. These fur seals li
on rough, rocky beaches on isolated islands in the souther
Atlantic and Indian Oceans (Map 133).

Arctocephalus townsendi —
the Guadalupe fur
seal

This is the only other fur seal found north of the equator. The
were hunted until it was thought they were extinct, but the
were rediscovered in 1954 on the Island of Guadalupe o
Southern California (Map 134). Some individuals may reach Sa
Miguel to the north in the California channel islands, an
Cedros Island, Baja California to the south. Adult males a
almost two metres long (six feet) and are dusky black though tl
head and shoulders appear greyish. Like the Juan Fernandez se
(see below) they have an extremely long pointed nose.

Arctocephalus philippi—
the Juan Fernandez
fur seal

Very little is known about this animal which was hunted for i
fur until the species was virtually extinct. It is similar to tl
Guadalupe seal, with a long pointed nose. They grow to tw
metres (six feet) and are blackish brown with a heavy mane
greyish colour. Both this animal and the Guadalupe fur seal ha
the habit of inverting themselves in the water, leaving their hir
flippers swaying above the surface. They are found in the Jua
Fernandez group of islands off the coast of Chile (Map 135).

Arctocephalus forsteri —
the New Zealand
fur seal

The New Zealand fur seal grows to just over two metres (s
feet) and is blackish grey with a browner belly. It feeds on squ
and fish and occasionally penguins and shags. The seals we

formerly more numerous and occurred all around the shores of New Zealand. They were part of the Maori diet, and were also eaten by Captain Cook. But the sealing industry reduced their numbers drastically — up to 400,000 skins being taken in two peak years. By 1840 their numbers were so low that they were not worth hunting. They now live on rocky islands around South Island, and they leave only for feeding expeditions (Map 136). They are also found on the coast of Western Australia.

Antarctic fur seal males grow to just under two metres (six feet); the females are about half a metre (two feet) smaller. The male is a grey brown with a greyish mane but the female is grey brown on the back with a white to grey belly or chest. This species is found on Antarctic islands around the Antarctic convergence (Map 137). This seal was hunted virtually to extinction but has now recovered. It is possible that the most recent increase in numbers is due to an abundance of krill — its food — caused by the decline of the baleen whales.

Arctocephalus gazella — the Antarctic or Kerguelen fur seal

This is a very little-known fur seal that lives only on the Galapagos Islands (Map 138). It is the smallest fur seal, growing to one and a half metres (five feet). It is a light tan colour with a greyish brown colouring down the back.

Arctocephalus galapagoensis — the Galapagos fur seal

There is only one species in this family but this is divided into two subspecies separated by size and geography.

Family 13: Odobenidae — walrus

Walrus are unmistakeable, having a square head, stiff whiskers, long tusks and a rough, wrinkled skin. Males grow to nearly 4 metres (12 feet); the females are slightly smaller. Both sexes have the tusks — really extra long canine teeth — which are about three centimetres long (one inch) at one year old but by eleven years may have grown to 100 centimetres (over three feet) and weigh over five kilograms (nearly twelve pounds). The breeding season is April to May, when the pups are born at over a metre (nearly four feet) long.

Odobenus rosmarus — the walrus

They may live for up to 30 years, with man and killer whales being their only enemies, although polar bears may take the occasional young. Most walrus migrate, moving south with the advancing ice in winter and returning north the following year. A few animals remain in the north in winter, when some may become seal-killers and will in turn be hunted by eskimos to protect their hunting grounds. Walrus forage in gravelly sea-beds at depths of 80 metres (260 feet), using their tusks to stir up the bottom and sometimes excavating long channels to dig up various mussel- and clam-type molluscs. They feed by sucking the animal out of its shell. They will also eat several different types of sea urchin, and fish.

Two subspecies of walrus are recognised, separated according to their size and location. (There may be a third subspecies between these two in the north of Russia.) These animals are found in shallow water around the Arctic coasts, usually staying close to the land or ice (Map 139). The Atlantic walrus, *Odobenus rosmarus rosmarus*, occurs on the Arctic coast of Canada around Baffin Island, on the west coast of Greenland, and around Iceland and northern Norway and Finland. It has been recorded rarely as far south as the Bay of Fundy, between New Brunswick and Nova Scotia in the west, and twice in Norway and on the Aberdeenshire coast in the east. The Pacific walrus, *Odobenus rosmarus divergens*, is slightly larger and occurs mainly in the Bering Sea and north-eastern Russia.

Family 14: Phocidae — the true seals

These animals have small front flippers and the back limbs cannot be turned forward. All the flippers are covered in fur though this is difficult to see when the animal is in the water. True seals have no external ear and are generally not as agile on land as the eared seals. There are four groups: the northern true seals, the Antarctic seals, the monk seals (the only species that lives in tropical waters) and the hooded and elephant seals.

Erignathus barbatus — the bearded seal

Bearded seals grow to about two and a quarter metres (seven feet). They are grey with a brown or reddish tinge on the head, and a dark brown band running down the back. The characteristic features, which give the animal its name, are the large whiskers on its face. Norwegian sealers called it the 'square flipper' because of the squared edge to the front flipper. The breeding season is normally April and May, but may be March in the south or June in the north. The pups are born on open ice floes with a brown woolly coat that is moulted when they are weaned.

Bearded seals whistle to their pups under water. They are curious and will be attracted to a boat, but if a hunter comes on them suddenly they seem to be struck rigid with fear and are easily harpooned. Man and polar bears are their only enemies. Eskimos use the seal hide for boot soles, heavy ropes, harnesses and kayak covering; the intestines are sometimes used for window panes. The seals feed on seabed animals such as crabs, shrimps, mussels, whelks and octopus. But they are not commercially important due to their scattered distribution.

These animals live around the world in the Arctic from the Gulf of Saint Lawrence to northern Norway and Alaska, although isolated individuals have been reported as far south as Scotland (Map 140). They prefer shallow coastal waters, free of ice, and may drift a little on ice floes in spring but they do not migrate.

Halichoerus grypus — the grey or Atlantic seal

Grey seals grow to just over two metres (over six feet); the females are about half a metre (18 inches) smaller. They have darker backs and lighter bellies and may sometimes have spots, but their colour varies with all shades of dark to light grey, brown and silver. Males have a darker background and lighter spots whereas the females have a lighter background with darker spots. Sometimes they have a growth of green 'sea grass' on their backs. Adult males have a high arched 'Roman nose' and sometimes a small external part of the ear is visible.

Mating often takes place in the water, depending on the territory, and the breeding season may last several months. A second moulting haul-out occurs in spring. They feed primarily on fish, crabs, shrimps and mussels. These seals are not killed commercially but they are sometimes the cause of complaints from fishermen arguing that they are damaging fish stock.

The grey seal lives on open rocky shores on both sides of the Atlantic including the British Isles (Map 141). They do not migrate, but there is a general dispersion at sea after breeding. The young spend their first two years at sea and may wander widely. The Baltic and Saint Lawrence populations breed in February and March; the British and European populations breed between September and December.

Phoca vitulina — the common, harbour or spotted seal

Adult male harbour seals grow to just under two metres (about six feet), and females are one and a half metres (five feet) long. They vary in colour but are basically light grey with irregular black spots and blotches. The belly is lighter. Generally, they are born on land between May and June, but the western Pacific animals breed earlier and the pups are born on ice floes between February and April. They eat fish, shrimps, squid, whelks, crabs and mussels. This species is also the cause of much aggravation from fishermen (see p. 141).

These seals are shore-living and are found mainly in estuaries and in places where sandbanks are uncovered at low tide. They occur on both sides of the Atlantic and Pacific, and several different subspecies have been recognised from different areas (Map 142).

Phoca hispida — the ringed seal or floe rat

Ringed seals grow to about one and a half metres (just under five feet). They are light grey with black spots, especially on the back, and many spots have lighter rings around them though these may merge in the middle of the back. Breeding takes place between March and April and they are born on land-fast ice — either in a lair under the snow or in a natural ice hollow. A breathing hole opens into the cavity so that the mother may approach the pup without being seen.

They feed on small fish and krill. Their enemies are killer whales, polar bears, Arctic foxes and man. They are hunted for fur and meat. Eskimos eat the flesh, and the skin is used for clothes — white pup skins often being used as underclothes. Ringed seals occur around the northern Arctic (Map 143). It is the commonest Arctic seal species and sometimes subspecies are described according to their area.

Phoca caspica — the Caspian seal

This seal is found in the northern Caspian Sea in winter but moves south in the summer. It grows to about one and a half metres (five feet) and is greyish yellow with irregular spots of brown or black, occasionally with light rings around the spots. The main food is small fish, shrimps and crabs. Pups are born on the ice in January or February.

Phoca sibirica — the Baikal seal

This animal is found only in the deepest lake in the world — Lake Baikal in eastern Russia. Both the Caspian and Baikal seals are probably the descendents of marine seals that were cut off millions of years ago during the great earth movements.

*Phoca groenlandica —
the Greenland, harp
seal or saddleback*

Harp seals grow to between one and a half and two metres and are a whitish grey with a black face and a large black horseshoe shape — the harp — running along the flanks and across the back. The female's face and harp are much paler, and may be broken up into spots. New-born pups have a coat of stiff white wool which is moulted after a month leaving a short-haired coat of grey with darker spots. The adults mate in the rough hummocky ice where they shelter the pups and keep a hole open into the sea. Males mate with one female and breeding may last from January to April. They feed on krill and small fish. These seals are found in open waters across the North Atlantic from Newfoundland to northern Norway and the Kara Sea (Map 144). They move north in summer and south in the spring to breed.

The white-coated pups are the main objective of seal hunters. They have to be killed at two to ten days old, though the young grey coat is also valuable. Immature seals or **bedlammers** (from the French *bête de la Mer* meaning 'beast of the sea') are taken for blubber, leather and oil. The killing of young harp seals has received much public attention, mainly due to the inhumane method of clubbing them to death: 180,000 harp seals are killed in this way each year for luxury items and it is possible that their future could be threatened.

*Phoco fasciata —
banded or ribbon seal*

Little is known about these seals. They grow to about one and a half metres (five feet) and have a distinctive pattern. The males are dark chocolate brown with white or yellow-white bands around the neck, round the rear of the body, and forming large circles around each front flipper. The females and young are paler. Breeding occurs from July to August. They eat fish, squid and small crustaceans. Banded seals are found in the north western corner of the Pacific (Map 145). They may travel south with the ice in winter and follow its retreat north in the summer.

*Leptonychotes weddelli —
the Weddell seal*

Weddell seal males are about two and a half metres long (just over eight feet), and the females are slightly larger. They have black topside, which often has white streaks and splashes, and a grey belly, but the colour fades during the summer to a rust brown. These seals are relatively free of natural enemies. During the winter they spend much of their time in the water and can be heard calling to each other below the ice as a warning to other seals to keep away from individual territories around their breathing holes. They keep breathing holes open in the ice by sawing and biting that wears down their teeth and often leads to abscesses. They normally eat fish, but squid and bottom-living animals are also eaten.

Weddell seals are the most southerly, living around Antarctic (Map 146), but they are occasionally seen near New Zealand, the Falkland Islands and Kerguelen Islands, and South Australia.

*Lobodon carcinophagus —
the crabeater or
white seal*

These seals grow to two and a half metres (eight feet) and the females are sometimes larger than the males. Young crabeaters are silvery brownish grey with chocolate brown markings on the sides and shoulders, shading to a pale belly. The coat fades during the year to a creamy white. In January this coat is moulted and replaced by a darker one, but the older animals get progressively paler.

Their main enemy is the killer whale and most adults bear the long parallel scars that tell the tale of a close encounter. Small scars around the neck are the result of fighting between males. They feed on krill, which they eat by swimming into a shoal with their mouths open to suck in the food. The water is pushed out through the specially adapted teeth to leave the food inside.

These are the most abundant seals of the Antarctic (Map 147) although exact population estimates are difficult; they may number about 15 million. They are found in and around the pack ice, moving north and south along with the ice. Stragglers have been recorded from Australia and New Zealand and even Uruguay.

Male leopard seals grow to about three metres (ten feet) and females to three and a half metres (about twelve feet). They are dark grey on the back, lighter on the belly, with dark and light grey spots on the throat, shoulders and sides. The long slim body and large head with a wide gape make the seal easy to recognise; it has been reported resembling a large snake.

*Hydrurga leptonyx —
the leopard seal*

Many have deep scars from killer whales which are probably their only enemy. They will eat fish, squid, penguins, bits of whale carcasses, and pups of other species. When a penguin is caught it is shaken so vigorously that the body is thrown out of its skin.

Leopard seals normally live on the edges of the pack ice, but they travel north to southern Australia, New Zealand and other sub-antarctic islands (Map 148).

The Ross seal grows to two and a half metres (eight feet). It is dark grey with streaks running back from the head, sides of the neck and shoulders. The body is plump, with large flippers, and the head can be withdrawn into rolls of fat around the neck. Squid, krill and fish are its main food. This seal is rarely seen as it usually occurs singly on the pack ice around the edge of Antarctica (Map 149).

*Ommatophoca rossi —
the Ross seal*

This seal grows to two and three-quarter metres (eight feet). It is chocolate brown although the hairs are tipped with yellow and the belly is greyish. Pups are born in September and October with black woolly coats. Little is known of the animal, though it probably feeds on fish.

*Monachus monachus —
the Mediterranean
monk seal*

In ancient Greece monk seal skins were thought to give protection against lightning, hailstones and gout; a flipper under the head at night gave relief from insomnia. But although this seal may at one time have been common in the Mediterranean and Black Seas, with the increased traffic it has retreated to less populous parts of its range, and there may now be only 500 to 1000 left (Map 150).

Monachus tropicalis —
the West Indian monk
seal

This seal was plentiful enough to be hunted for oil in the past, but today it is believed the animal is extinct. Its range was the Bahamas, Guadeloupe and the Caribbean Islands.

Monachus schauinslandi —
the Laysan or
Hawaiian monk seal

These seals spend their time in shallow water and on sandy beaches in the sun; they are mostly nocturnal. Females are slightly larger and much heavier than the males which may grow to two and a quarter metres (seven feet). Pups are born chiefly around March to May, and have a coat of soft black hair which is moulted to a silvery blue grey, shading to light underneath. Adults moult between May and November. They have scars that are probably caused by the coral of their island environment. They feed on eels, reef fish and octopus. This seal lives on the atolls and islands of the Leeward chain to the north-west of the main Hawaiian Islands (Map 151). Stragglers have been noted as far as Hawaii. Although once numerous, they have been much depleted due to sealing expeditions.

Cystophora cristata —
the hooded, crested or
bladder-nose seal

Hooded seal males grow to two and a half metres (eight feet): the females are about two metres long. They are grey and covered with irregularly shaped black patches. Adult males have an enlarged nasal cavity that forms an inflatable hood on the top of the head. Normally it hangs down in front of the mouth, but when inflated it forms a high cushion on the head, twice the size of a football. It is possible that they inflate this hood in time of anger in the wild but its function is not really clear. Pups, born at the end of March or beginning of April, are called blue-backs in the fur trade due to their silver blue colour; their greatest enemy is man. The seals eat octopus, squid, shrimps and fish besides starfish and mussels from the seabed. They are found from Newfoundland, across the North Atlantic to Spitsbergen and Bear Island and on to the Denmark Strait (Map 152). They prefer deep water and thick ice floes but stragglers have been found as far south as Florida, the Bay of Biscay, and Portugal.

Mirounga leonina —
the southern elephant
seal

This is the largest seal: males grow to 6 metres (up to 20 feet), females to 3.5 metres (12 feet). The male is dark grey and a little lighter on the belly, but fighting leads to intense scarring on the neck, and this scar tissue makes the skin of the neck and chest extremely thick. Females are browner and dark, and often have scars on the neck from bites inflicted during mating.

The males have an inflatable proboscis or snout, the tip of which overhangs the mouth. It is erected during the breeding season and may act as a resonating chamber to enhance the roar that may be heard for several miles. Breeding starts at the beginning of September, when the male gathers a large harem of up to 100 cows. At birth the pups have a coat of black woolly hair that moults after about ten days. Adults return to land between December and February to moult.

They feed on fish and squid. Their only enemies are the leopard seal, which takes pups, and the killer whale. They were intensively hunted by man, especially with the diminished numbers of fur seals. They are found around the southern oceans on most of the subantarctic islands and even as far north as

Mauritius and, very rarely, Saint Helena (Map 153).

These seals are similar in size and colour to the southern species, though they appear more placid and rarely rear up into the U-shape so often seen in the aggressive southern animal. The inflatable proboscis is larger in northern seals and may hang down in front of the mouth by 30 centimetres (twelve inches). When erected it has a groove that almost cuts it in two. Their general breeding behaviour is similar to that of the southern species, but the harems are smaller — usually consisting of about a dozen females. They feed on squid, fish and sometimes small sharks, taking food from depths of 100 metres (330 feet).

Mirounga angustirostris — the norther elephant seal

The populations show a remarkable recovery from the near extinction caused by the sealing industry. They were finally protected in 1922. They are found off the coast of Southern California and Mexico although they have been recorded as far north as Prince of Wales Island in Alaska (Map 154).

SEALS AND HUNTING

Seals have been hunted for their skin, meat and fur for many years with such intensity that many are now protected to prevent them from becoming extinct. But one question that has received much public attention recently is the annual seal culls that fishermen claim are necessary because the seals have reached such numbers that they are reducing fish stocks and damaging equipment. The arguments are charged with emotion, the fishermen claiming to be protecting their livelihood, while the conservationists claim that the seals are threatened.

The essential difficulty is that there is insufficient objective information about the number of seals, their behaviour, and their effect on the fishing industry. It may be that increased fishing activity by the fishermen has put them in competition with the seals. It is also a fact that a fishing community may regard a particular species as being too numerous when in fact their particular area may be one of the few breeding sites and taking the world-wide view would indicate that the total numbers are low. A very important point is that so little is known about the natural changes in populations of both seals and fish that many other factors could also be involved. Much more detailed work is required about these natural changes if the problems are to be solved quietly, calmly and efficiently.

ORDER 3: SIRENIA — SEA COWS

It is thought that these animals may be the origin of stories about mermaids because they recline in a semi-upright position when suckling their young. They are named after the mythological sirens or sea nymphs that were thought to lure sailors onto treacherous rocks with their songs. The animals are large, with paddle-like flippers, a flattened square tail and a blunt muzzle with stiff hairs. They are vegetarian, eating up to 10 kilograms (22 pounds) of food a day. The animals have suffered greatly from the hunting and sporting activities of man.

Family 15:
Dugongidae —
the dugongs
Dugong dugong —
the dugong

These animals grow to three metres (ten feet) and are blue grey above and paler below. They are to be found from the Red Sea to Madagascar and around the Far East south to Australia (Map 155). But they are rare.

Family 16:
Trichechidae —
the manatees
Trichetus manatus —
North American
manatee

This species lives on the coast and in the coastal rivers of the south-eastern United States, especially Florida, and in the Caribbean and along the north-eastern coast of South America (Map 156). They grow to three metres (ten feet) and are a dull grey colour. Like other manatees this animal is hunted for food, but it also suffers from collisions with boats and from losses during periods of cold weather.

Trichetus inunguis —
the South American
manatee

The South American manatee is restricted to fresh water and prefers the slow-moving or still waters in the Amazon river system (Map 157). It grows to two metres (six feet) and has a white breast patch and elongated flippers. It is now extremely rare.

Trichetus senegalensis —
the West African
manatee

This animal is very similar to the North American manatee. It lives in tropical West Africa from Senegal to Angola, although it is not restricted to the coast and will travel hundreds of miles up rivers (Map 158). Its meat is valuable, which has led to hunting and a consequent decline in numbers. But, for the present, in some regions it remains relatively plentiful.

ORDER 4: CARNIVORA — SEA OTTERS

Family 17:
Mustelidae — otters

Although this family includes bears, badgers, weasels and their allies, only the sea otter is truly marine although some species of river otter, *Lutra* spp. especially the chungungo *Lutra felina* of Chile and Peru feed extensively in the sea.

Enhydra lutris —
the sea otter

Sea otters feed on mussels, abalones and sea urchins from the shallow sea-bed, picking up their prey and eating it lying on their backs on the surface (Colour Plate 60). They formerly ranged from Baja California, north to Alaska and throughout the Aleutian, Pribilof and Commander Islands, along the south-east coast of Kamchatka and through the Kuril Islands to northern Hokaido. But their pelts were much sought-after, and the animal was hunted virtually to extinction in some parts of the world.

Now they are protected and appear to be recovering. They spend most of their lives in the water among the kelp seaweeds but they breed on land.

There are three subspecies: *Enhydra lutris nereis* lives from Mexico to Prince William Sound, Alaska; *Enhydra lutris lutris* is found in the Aleutian and Commander Islands; *Enhydra lutris gracilis* is found in Kamchatka and the Kuril Islands (Map 159). The Aleutian race was introduced into Oregon, Washington, British Columbia and south-eastern Alaska to replace animals that were hunted to extinction.

11. The Sargassum Community
The Animals on Gulfweed

There are many tales of old sailing ships becalmed for days in areas of weed. Seafarers have cursed the weed, blaming it and the lack of wind for their slow progress. Some of the more imaginative seamen added spice to their stories by telling of strange sea serpents and animals living in the weed.

Today, merchant ships take little effort to sail through these once-feared areas, but the weed is still there and, although there are no sea serpents — at least not in the Atlantic — the weed is alive with animals, some creeping and crawling, others climbing, and some just sticking where they are. Together they make up a thriving community that is quite unique, drifting around the ocean's surface, miles from the nearest land. But it is a precarious existence. The weed is a life-raft; if an animal loses its grip it will sink to the depths of the ocean. Without the weed the community would perish, hence the animals are discussed collectively in this chapter although some species have already been referred to.

The *Sargassum* spp. belong to the same group as many of the brown seaweeds that are found on the sea shore; the name is derived from the area in the central Atlantic — the Sargasso Sea — where much of the weed is found (Map 160). It is also sometimes referred to as gulfweed, especially in and around the Gulf of Mexico. There are several species; some are found in the area of the Red Sea, others in the Pacific Ocean around Japan, but the animals that live among the weed are similar in all these areas.

Picking up clumps of the weed is a relatively simple task from a small boat but it can be difficult from a big ship, even if it is stopped. I have found that deft handling of a simple grapnel swung on a rope can catch a good sized clump of weed and, although some of the more active animals are lost, there are usually plenty left clinging to keep an enthusiast occupied for hours.

It was thought that the *Sargassum* originated from the shores and shallow water of the Caribbean, northern Brazil and Florida, but only a small amount of the weed starts in this way. Most of the *Sargassum* is the product of its own growth and development while drifting at the surface — this is its habitat.

The weed itself is fascinating. It is a tangle of 'stems', branches and leaf-like pieces with small round air bladders to keep it afloat. *Sargasso* is from the Portuguese word for grapes (Colour Plate 1). As with all living plants, it requires sunlight and must remain near the sea's surface.

The young growing tips of the *Sargassum* produce a chemical which is toxic and acts like an antifoulant, stopping the animals from growing on the tips. Therefore most of the community is found among the older branches of the weed. If clumps of *Sargassum* are collected, the water must be changed ever few hours if the animals are to stay alive; they are killed overnight if left in a bucket with the weed.

THE *SARGASSUM* COMMUNITY 'TREE'

There is a great variety of animals on the weed. Some are members o
groups that have already been discussed, but others belong to new
groups (Figure 11.1).

Many of these animals are similar to those found crawling on wee
and rocks on the sea shore; they can only live at sea by hitching a rid
on the weed.

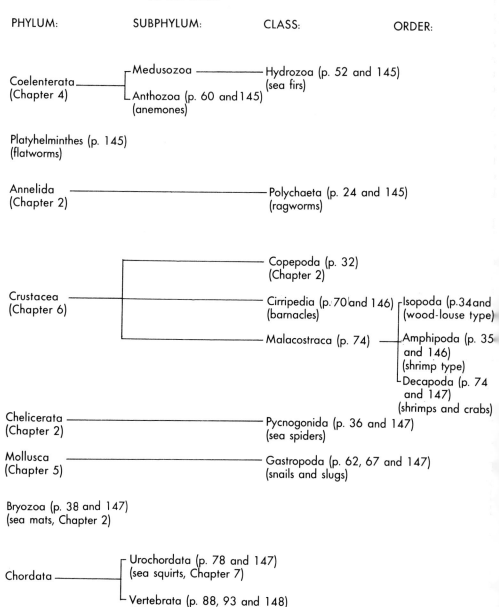

Figure 11.1: *The Sargassum community classification, or 'family tree'*

PHYLUM 1: COELENTERATES — SEA ANEMONES AND SEA FIRS (Chapter 3)

Two types of coelenterates are very common on *Sargassum*: hydroids or sea firs, and one species of anemone.

CLASS: HYDROZOANS — SEA FIRS

Sea firs are colonies of tiny animals living inside tubes growing, like runners with many branches, throughout the weed. There are anemone-like structures, with tentacles to catch food, inside little cups at the ends of some of the branches. In some species the cups are on single branches, but in others there may be many cups on branches arranged like a Christmas tree. Occasionally it may be possible to see the reproductive structures which are small jar-shaped containers with the young stages inside.

SUBPHYLUM: ANTHOZOA — SEA ANEMONES

Anemonia sargassensis, the *Sargassum* anemone, is cream to brown in colour and grows to between two and five millimetres long and five to ten millimetres in diameter — depending on how much it is contracted. There are 32 slender tentacles used to catch tiny planktonic animals for food.

PHYLUM 2: PLATYHELMINTHES — FLATWORMS

These simple little worms look like flattened pieces of tissue crawling rapidly on an invisible cushion, rather like a hovercraft. The 'cushion' consists of thousands of tiny hairs which carry the animal along. However, they can also swim by rapid undulations of the body. The mouth is underneath and the eyes are on the back of the head. There are four species associated with *Sargassum*, all about five millimetres long and coloured brown.

Flatworms have male and female sex cells in the same animal (hermaphrodite) but they fertilise each other and the eggs are usually laid in jelly-like clumps.

PHYLUM 3: ANNELIDS — THE TRUE WORMS (Chapter 2)

CLASS: POLYCHAETA — THE RAGWORMS

The ragworms have whiskers and swimming paddles down each side of the body (Figure 2.9 and Black and White Plate 4) — hence their common name. Five species can be found on *Sargassum* but the following three are the most common.

This worm grows to between three and twelve centimetres long and is brown, with rings running round the body. In the mouth are a pair of strong jaws for catching small animals and tearing up food.

Platynereis dumerili — the ragworm

Spirorbis is not immediately recognisable as a ragworm because it lives inside a little white coiled tube built from limestone. The worm feeds by pushing its head and tentacles out of the entrance to the tube to filter food from the water. if it is

Spirorbis corrugatus — the coiled tubeworm

disturbed it shoots back into the tube and closes the entrance with a 'plug'.

Amphinome rostrata —
the fire worm

This species is rare but is worth a note because it belongs to a family which have bristles that can cause a stinging irritation if not handled carefully. This particular species grows to 40 centimetres long and is bluish brown or grey with deep brown or red and orange tipped bristles.

PHYLUM 4: MOLLUSCS — SHELLFISH AND SNAILS (Chapter 5)

CLASS: GASTROPODA — SNAILS AND SLUGS

Gastropods are the only molluscs found on *Sargassum*. *Litiopa melanostoma* is the only snail that is a permanent resident, growing to about half a centimetre long with a brown coiled shell. Occasionally, the violet bubble-raft snail, *Janthina* spp. (Figure 5.3 and Black and White Plate 12) may be found, and sometimes when the weed is fairly near the coast, it may pick up some coastal species. But neither *Janthina* nor these coastal visitors are permanent members of the community.

There are six species of sea slugs in the weed, the most common being *Scyllaea pelagica* which may reach six centimetres in length. It has three pairs of ragged strap-like flaps that resemble the leaf-like pieces of weed. The oceanic sea slug, *Glaucus atlanticus* (Colour Plate 6) like *Janthina*, may also be a temporary visitor.

PHYLUM 5: CRUSTACEANS (Chapter 6)

Barnacles, shrimps and crabs are common members of the *Sargassum* community, but there are also several species belonging to the planktonic copepods which can be recognised by their characteristic shape (Figure 2.14).

CLASS: CIRRIPEDES — BARNACLES

Barnacles will attach and grow on many objects, and *Sargassum* is no exception with six different species of goose barnacle in the community (Colour Plate 37).

There are several *Lepas* spp., but the commonest is *Lepas pectinata* which grows to two centimetres long and has rays or grooves and occasionally spines on the limestone plates.

Conchoderma virgatum is an easily recognisable barnacle because the plates are not obvious, and the body is brown with cream stripe (Colour Plate 38).

ORDER: ISOPODS — WOODLOUSE TYPE

Four species are found in *Sargassum* but the most common, *Bargetus minutus* is two millimetres long and looks like a miniature woodlouse with a yellow/white back and brown spots.

ORDER: AMPHIPODS — SHRIMP-LIKE

About half a dozen species of amphipod are usually found in the weed, and three belong to the Caprellid family — commonly called ghost or skeleton shrimps because they are transparent and difficult to see. They look like little stick insects, hanging onto a branch — usually on a sea fir — with their back legs, and wafting their pincers in the water.

ORDER: DECAPODS — CRABS AND TRUE SHRIMPS

Leander tenuicornis is the largest *Sargassum* shrimp, growing to five centimetres long with a stout body and colouring to match the weed. The next in size is *Latreutas fucorum*, growing to two centimetres, and variably coloured. Both these shrimps are very common, but another species, *Hippolyte coerulescens*, which is only three millimetres long, may be found occasionally.

The *Sargassum* crab, *Planes minutus* is pea-sized (about two centimetres across) and light yellow to brown, with large heavy pincers and hairy legs. Other species, especially the swimming crabs of the Portunid family, and their young stages, are often found in the weed but they are not permanent community members — with one exception: the swimming crab *Portunus sayi* is a resident. It grows to three centimetres long by five and a half centimetres across and has the characteristic flattened swimming legs (Colour Plate 42).

PHYLUM 6: CHELICERATES — SPIDERS AND HORSESHOE CRABS

Horseshoe or king crabs, sea spiders and a few mites are the only marine animals in this group (Chapter 2).

CLASS: PYCNOGONIDS — SEA SPIDERS

Sea spiders have a thin segmented body with long spindly legs — rather like a 'daddy-long-legs' (Black and White Plate 7). They are active hunters, with sharp jaws at the end of a snout for devouring any smaller animals they catch. Sexes are separate and, after fertilisation, the male carries the eggs on his first two legs until they hatch as young sea spiders.

Three species are usually found crawling among the *Sargassum*. The largest is *Endeis spinosa*, growing to about three centimetres long, with legs twice the length of the body. Another very common species, *Anoplodactylus petiolatus*, is only about half a centimetre long and the legs have few hairs.

PHYLUM 7: BRYOZOA — SEA MATS

Growing in patches on the weed are lace-like encrustations built up of little boxes. These coral-like growths are colonies, with each individual animal living in a box. A small 'trapdoor' is located near the top of the box, through which the animal extends tentacle-like fingers to take food from the water. The most common species is *Membranipora tuberculata* in which each of the boxes has two small horn-like projections.

PHYLUM 8: CHORDATES

SUBPHYLUM: UROCHORDATES — SEA SQUIRTS (Chapter 7)

One species of sea squirt may occasionally be found living on the weed; it is a colony of tiny individuals buried in a jelly-like lump.

SUBPHYLUM: VERTEBRATES — FISH (Chapter 8)

Several species of fish use the *Sargassum* weed for shelter and hiding such as the dolphin fish, *Coryphaena* spp., and members of the Carangid family like the pilot fish, *Naucrates ductor*. Some flying fish lay their eggs in nests built from the weed. But two species of fish live permanently among the weed. One is the *Sargassum* fish, *Histrio histric* (Colour Plate 21). The other is the pipefish called *Syngnathus pelagicus*, which may be mistaken for a snake.

These fish are well camouflaged: the *Sargassum* fish is mottled yellow, brown and white; the pipefish is greyish brown, having speckled or indistinct white crossbands with black margins. This fish also has a dark longitudinal stripe in front of each eye, a cylindrical body up to 17 centimetres (nearly 8 inches) long, pectoral and dorsal fins, and a tail fin that ends in a point. The male pipefish carries the fertilised eggs in a brood pouch that runs along the belly and opens to the outside world via a long slit.

12. The Seashore and Marine Fouling

The Life on Harbour Walls, Pier Piles, Ships' Bottoms and in Coolers, Condensers and Filters

For as long as ships have sailed the oceans, animals and plants have grown on them and in them. For almost the same length of time seafarers have tried to prevent this fouling, and today there are antifouling paints which contain poisons to try and stop these organisms from growing on ships' bottoms. The paints have varying degrees of success but they are all limited, and occasionally ships must be taken out of the water, scraped clean and repainted.

The engine room is plagued by animals growing in coolers, condensers, seawater intakes and filters, all of which must be opened and cleaned every few months. Buoys, harbour walls and pier piles also collect luxuriant growths of seaweed and animals.

This fouling community represents a collection of organisms that will grow on virtually anything left in the sea. Many are common on seashores, which is virtually where the story begins.

THE SEASHORE

The shore is the area of sea-bed that at some time each day is not covered by the sea, due to the rise and fall of the tides (Figure 12.1). There are different types of shore — mud, sand, shingle and rock — each with its own particular group of species. Most of the life associated with ships and boats are usually found on rock, so it is with these animals and plants that we start.

The shore is unique because the organisms living here must be able to cope with being exposed to air, sunlight, rain and snow. At the top of the shore is the high-water mark — the highest point that the sea will reach. The bottom of the shore is the low-water mark; the sea rarely falls below this point.

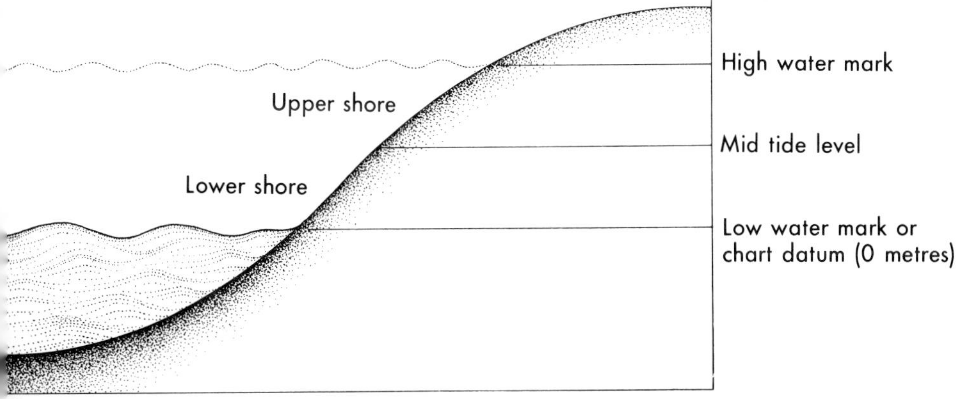

Figure 12.1: Zones of a simple sea shore

High water mark

Upper shore

Mid tide level

Lower shore

Low water mark or chart datum (0 metres)

The animals and plants living near the top of the shore are uncovered by the sea for longer periods than those nearer the bottom. The top-shore organisms must, therefore, be better adapted to living on the shore than those lower down. Some upper shore animals, for instance, have evolved to be air-breathing, so they can stay out of the sea for long periods: they are more like land animals.

water marks; the level of the water is different every day, and some organisms high on the shore may be left out of the water for several days at a time. This daily rise and fall of the tide is familiar to most people but the forces that cause tides are complicated. Tides vary throughout the world, on Atlantic coasts we are accustomed to two tides every 24 hours, but in some parts of the world there is only one and in others there is very little tidal movement. The tidal range also varies, for instance around Britain it is usually between three and five metres (10 to 15 feet), but in the Bay of Fundy it is around 16 metres (50 feet).

HOW TIDES ARE CAUSED

A detailed explanation of all the variations is beyond the scope of this book, but a simple understanding of the two high and two low tides every 24 hours can be obtained by considering the relationship of the earth, moon and sun.

All bodies experience the force of gravity and the amount of force depends on the sizes of the bodies and their distance apart. The bigger the body, the greater the force, but the further away the body, the less the force. The earth's gravity causes things to fall to the ground, and this gravity pulls on the moon. The moon also has gravity and is attracting the earth but this force is much smaller (about one sixth that of the earth) because the moon is smaller. If the earth and moon are pulling each other, why do they not crash together?

The moon is spinning around the earth, taking 28 days to do so — a lunar month. But the earth is also spinning around the moon. The two bodies are spinning around each other like two balls on the ends of a piece of string. If the earth and moon were equal they would spin around the midpoint between the two, but because the earth is larger they spin around a point inside the earth. The spin of the earth and moon around each other gives rise to a centrifugal force which tends to throw them apart — like swinging a ball on a string around you. If there were no gravity the moon would fly off into space — just as the ball would fly off if you let go of the string.

The gravitational pull of the moon has an effect on the earth. The effect on land is extremely small and can only be measured using very sophisticated instruments. But the effect on the oceans is much more noticeable. This effect is easier to understand if we imagine that the earth is covered in water (in fact 70 per cent of the earth's surface i under water). The area of water directly under the moon is pulled into a bulge. This is high water (Figure 12.2).

At the opposite side of the earth another bulge of water is formed due to the centrifugal force of the earth spinning around the moon. If you were standing at A or B you would be at high tide, but at C or D it would be low tide because the water has been pulled away to form the high tides at A and B.

But the earth revolves once every 24 hours, so if you are standing at A at midnight, in six hours you will have moved to C and low water. At midday you will be at B and high water again, and by six in the evening at D and low water. By midnight you will be back at the start — well almost.

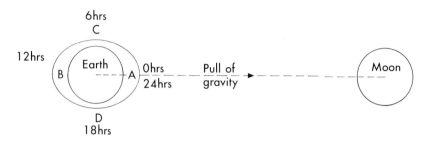

During the last 24 hours, the moon has moved part of the way on its orbit round the earth, so it actually takes 24 hours and 50 minutes to get back directly under the moon and at high water again. High tide will now be at 12.50 in the morning (Figure 12.3). This explains why the tides do not occur at the same time every day.

Figure 12.3: After twenty-four hours the moon has moved on its own orbit (the earth is viewed from the north pole)

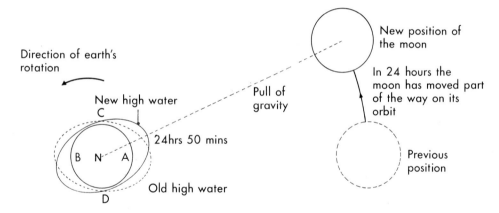

DIFFERENT KINDS OF TIDES

To explain the different heights that the tide reaches every day we must bring the sun into the picture. The sun is much larger than the earth or moon and therefore has a greater force of gravity, but the sun is much further away from the earth than the moon so the force is less — about two-thirds that of the moon. Figure 12.4 shows the sun in relation to the earth and the moon.

Figure 12.4: The effect of the sun on the tides

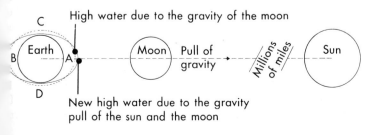

In this figure the sun is adding to the pull of the moon and the water is pulled into a bigger bulge giving an extra high tide at A. An extra bulge is also formed at the opposite side of the earth, and more water is pulled from C and D which leaves these points with extra low waters. These extra high and low tides are called **spring tides**. There is no connection between the season and a spring tide; the name comes from Nordic mythology that tells of a spring beneath the oceans which tops up the water level.

The moon orbits around the earth about once a month — in fact every 28 days. After seven days the moon is at a right angle to the line of the sun (Figure 12.5) and is seen on earth as a half-moon. The water is being pulled one way by the moon and another way by the sun with the result that the high water is not so high and the low water is not so low. These are **neap tides**.

Figure 12.5: The effect on the tides of the moon's orbit round the earth

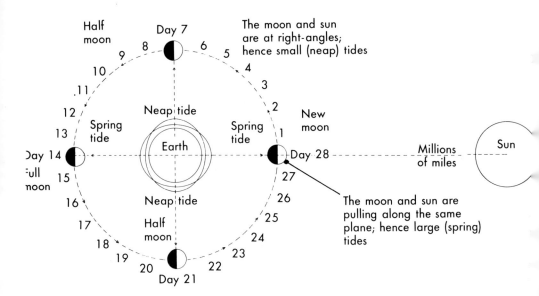

After seven more days the moon is round the other side of the earth and is seen as a full moon. The sun and moon, although on opposite sides are again pulling in the same plane. This gives another set of extra high high waters and extra low low waters which are spring tides once again.

On day 21 the moon is again pulling at right-angles to the sun so high water is lower, and low water is higher — so once again these are neap tides.

The picture is complicated but basically at a new moon (when no moon is visible on earth) and full moon there are large spring tides with the highest high waters and lowest low waters. Between these times the rise and fall of the tides gets less until at the half moon stages there are lower high waters and higher low waters — neap tides.

OTHER INFLUENCES

Many factors can effect the tides, but the main one is the elliptical orbit of the earth and moon around the sun (Figure 12.6) which takes a year, or more accurately 365¼ days. The earth and moon are nearest the sun in March and September so the effect of the sun's gravity is greater and the tides are bigger. In June and December, when the earth is farther away, the sun's effect is less and the tides are smaller.

Figure 12.6: The elliptical orbit of the earth and moon round the sun

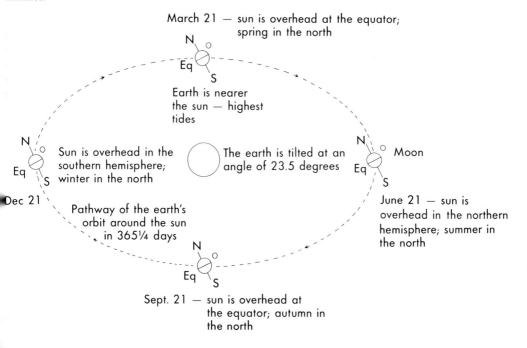

March 21 — sun is overhead at the equator; spring in the north

Earth is nearer the sun — highest tides

Sun is overhead in the southern hemisphere; winter in the north

The earth is tilted at an angle of 23.5 degrees

Moon

Dec 21

Pathway of the earth's orbit around the sun in 365¼ days

June 21 — sun is overhead in the northern hemisphere; summer in the north

Sept. 21 — sun is overhead at the equator; autumn in the north

SUMMARY

There are two tides every 24 hours.

The amount of rise and fall of the water changes every day. In one month the largest tides (spring tides) are at new and full moons; the smallest tides (neap tides) are at the half moons.

March and September have the largest tides in the year; June and December the lowest.

But the picture in the real world is much more complicated. The moon does not orbit the earth is such a simple way and the tides are not formed in an ocean-covered earth but the water is contained in huge basins of various sizes. Tidal theory is a complex and fascinating subject and will repay further reading.

THE SHORE

The different tides affect the shore, which can be divided into three areas according to the level of the water (Figure 12.7).

Most of the shore will be covered and uncovered by the tide twice roughly every twelve hours, but the top of the shore may be left exposed for up to three weeks, and even when it is covered it may be only for an hour or so. During high tide the conditions over the whole shore are fairly constant, but this changes as the tide falls. In hot weather the temperature will increase; in very cold weather the temperature will fall. These changes affect the animals and plants.

Changes in temperature also affect the amount of oxygen that is dissolved in the water in rock pools. In hot weather water will evaporate from pools causing the salinity to increase, or it might rain and so the salinity will fall.

Figure 12.7: The sea shore tidal marks

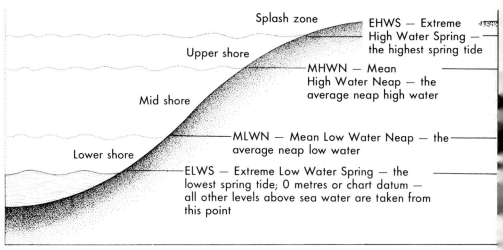

The animals and plants living on the shore must be able to cope with these changing conditions. On exposed surfaces as the water evaporates there is a danger of dehydration and animals such as barnacles, mussels and snails close themselves off from the outside and wait for the returning tide. Those species living at the top of the shore must be able to tolerate being exposed for long periods. Some of these animals have even developed simple lungs to breathe air, but there are very few species that can withstand these extreme conditions. Here animals such as the grape-pip periwinkle hide in cracks while lichens grow on the open surfaces. In the middle shore are those species that need to be covered with seawater at least twice every day. There are more species here such as barnacles, marine snails, seaweeds, etc. The lower shore may not be uncovered by the tide at all for certain periods; the species living here are least adapted to conditions on the shore. Large strap-like seaweeds (kelp) grow in this area and there is a rich variety of animals including anemones fish, crabs, sponges, sea squirts and so on.

The level that the tide reaches therefore determines which species live in different areas or zones. This zonation of species can be used to determine which parts are the upper, middle and lower shores. The upper shore is the periwinkle/lichen zone, the middle shore is the barnacle/seaweed zone and the lower shore is the kelp zone.

Sand and mud shores are also affected by the different levels of the tides and the species live in separate zones but this is not as obvious as on a rocky shore because most animals live buried in the sand and there are very few seaweeds.

Other factors affect the shore life such as whether the shore is facing the sun or not; this controls how quickly it will dry up when the tide falls. Animals and plants living on a slope facing the sun must be able to resist drying out, those species without this ability will live on a slope sheltered from the sun. Cracks and holes in the surface will also provide shelter from the sun and rain for some species. The amount of pollution will also affect the life on a shore.

The shore is also affected by waves. Some are exposed to a constant battering of oceanic waves while others, in sheltered bays or inlets, have small, gentle waves. There is a range of exposure between these two extremes. On an exposed shore the animals and plants must survive the strong crashing waves. Such a shore has very few seaweeds which would be torn away and only those firmly attached animals such as barnacles, limpets and mussels can survive. In the same area with the same tidal ranges a shore exposed to large waves is wider than one that is sheltered because the wind and waves extend the influence of the seawater further up the shore. In a sheltered bay or inlet the shores are covered in seaweeds that provide shelter and damp places for many animals such as crabs, fish, sea anemones and sponges. It is possible to tell roughly how much a shore is exposed by looking at the types of animals and plants living there.

Eventually all these influences can be put together to give a complete picture of the life on the sea shore. It is a constantly changing place with the organisms not only battling with each other but also with the elements — a difficult place for marine organisms to live.

MARINE FOULING

This is the term given to the undesirable growth of animals and plants on ships, buoys and other structures. There are four and a half thousand fouling species ranging from simple bacteria to seaweeds, worms, barnacles, and anemones. The most common are the seaweeds and barnacles.

A jetty built on piles extends out from the shore and is often in the main stream of water and therefore fairly exposed. The piles are usually covered in barnacles, with bunches of mussels clustered round the midshore to low-water mark. It is not advisable, in many areas, to eat the mussels because of the probability of industrial and sewage contamination of the water. A Frenchman in Le Havre was very concerned about my health when he saw me picking mussels off the jetty, although I did not intend eating them. Crawling among these animals are small crabs and ragworms, and there may be the occasional firmly attached sea anemone such as the feather anemone, *Metridium senile* (Colour Plate 62) which is common in temperate waters.

Harbours built inside breakers are more sheltered and will have more seaweed growing among the ever-present barnacles. It may be possible to see the difference due to exposure to waves by viewing either side of a harbour wall and comparing the seaward face with the sheltered landward side.

Today, many jetties are built from metal and concrete, but there are still a few wooden structures around. Not only do animals grow on these but some burrow into the wood and can weaken the piles. Two commercially important wood-borers are the gribble and the shipworms.

Limnoria spp. — the gribble

The gribble is a crustacean that looks like a tiny woodlouse (it is an isopod; see Chapter 2). It lives by eating its way through the wood, excavating long tunnels and occasionally making holes in the surface to provide fresh water for respiration (Figure 12.8).

Figure 12.8: A section through a piece of wood attacked by the Gribble — Limnaria *sp.*

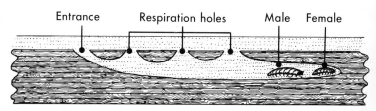

Teredo sp. and *Bankia* sp. — shipworms

After Christopher Columbus abandoned his two ships *Capitana* and *Santiago* in Jamaica he wrote: 'The ships are more perforated with worms than a honeycomb.' These holes were caused by molluscs that have a long body, a small shell at the front end and two long tubes called siphons projecting from the back. They live in burrows in the wood with the siphons peeping out of a small hole (Figure 12.9). As the animal grows the burrow is extended and lined with a calcareous (like limestone) deposit so the wood may feel deceptively sound. The tunnels may reach between 18 centimetres and 2 metres depending on the species. The wood is cut by opening and rocking movements of the shell and the resulting sawdust is used for food. They also feed on tiny plankton sucked in through the inlet siphon. The most common shipworms are *Teredo* and *Bankia*.

Figure 12.9: A section through a piece of wood attacked by the Shipworm — Teredo *sp.*

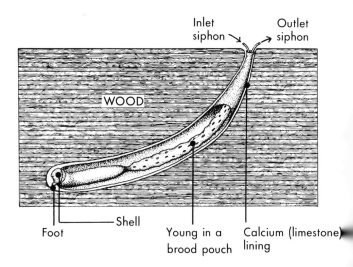

Many species are found in the engine room: they include barnacles, tubeworms, crabs, snails and occasionally jellyfish. Today a well-maintained hull is fairly clean, but Colour Plate 6 shows the extent to which fouling can develop when a vessel is not painted regularly. A jungle of seagrass, *Enteromorpha* spp. will soon grow on a dirty hull, which then provides shelter for tubeworms, barnacles and even sea squirts.

MARINE ANTIFOULING

In the eighteenth and nineteenth centuries, copper sheathing was widely used to protect boat hulls, although it was not until 1824 that Sir Humphrey Davy showed that it was the slowly dissolving copper that prevented the attachment of organisms.

That is still the basic idea today — with some refinements. A toxic compound is incorporated into a paint to be applied to the ship. The hull is thus coated with a thin toxic layer of seawater which lasts until most of the compound is exhausted; then it is time for a repaint. Copper is still used a great deal, often in combination with other poisons such as mercury, arsenic, tin, lead and zinc. However, the use of all these metals in the marine environment is causing concern.

Research is continuing in an effort to find new and better ways of preventing fouling. There are many compounds and combinations, some more effective than others. One is incorporated into a rubber matrix, and it is claimed that this paint gets smoother as the toxic compound dissolves away. There are even some paints that can be applied underwater. The battle continues against this simple but very effective group of resilient animals and plants.

13. The Future of the Marine World

The oceans provide three main services: a store of material resources, a means of transport and a complex of non-material uses. These activities can conflict with each other and with the need for the conservation of marine life.

1. Material Resources. (a) oil and gas exploration; (b) sand and gravel extraction; (c) waste disposal and (d) fisheries. These uses can conflict. For instance, the establishment of permanent structures excludes any fishing; there is the continuous threat of oil pollution from blowouts and/or fractured pipelines; sand and gravel extraction damages the sea-bed and could affect fish nurseries; waste disposal creates a pollution problem.

2. Transport. Thousands of ships carry goods and people around the world. In areas where shipping is concentrated, such as in the North Sea, there are navigational hazards. Shipping operations such as chemical and oil tanker washings and the dumping of waste provide a source of pollution.

3. Non-material Resources. (a) recreation; (b) research and education; (c) conservation and (d) defence. To varying degrees (a), (b) and (c) involve an appreciation of the marine environment. The quality of this environment is affected by interactions between and within the transport and material resource users. There is also conflict between these 'quality' users. Recreation creates tourist pressure — boating, SCUBA diving and fishing; research and education may be responsible for the depletion of some species by collecting and constant use. The military use the sea for exercises, as a dumping ground for explosives and in some cases as a front line.

FOOD RESOURCES

The life of the oceans could be a renewable resource of protein-rich food for the world's increasing human population, so why is the fishing industry declining in so many areas? The reasons are obvious in some cases, but not so in others because we do not know enough about natural changes in the numbers of marine organisms. Nor do we know enough about changes in the climate and current systems that affect marine life. In 1972, the Peruvian anchovy fishery collapsed due to the occurrence of *El Niño*. This is a change in the wind direction that causes drastic changes in the currents. Warm equatorial waters penetrate much farther south than usual and kill the anchovies; consequently many other animals that feed on the fish either die or leave the area. This collapse has repercussions world-wide — and was partly responsible for an increase in the world price of grain.

Fishing is a notoriously risky business because there is considerable variation in the amount and type of fish available. Over the past 50 years the fishing effort has intensified throughout the world and today it is becoming less selective, partly because of the increased tendency to use the fish for animal feed and partly because of the falling numbers of the popular species of fish. There is a limit to the amount of fish that can be taken year after year from each natural stock of fish, and today's 'efficient' fishing industry can develop from small beginnings to over-utilisation in just a few years. But the fishing fleets cannot continue indefinitely moving from over-fished areas to other regions. Most of the world's fish stocks are now fully, or almost fully, exploited (Figure 13.1).

More efficient fishing methods do not necessarily mean an increase in efficiency for the whole industry. To make fishing more effective in the long term, we need more information about fish growth and development; about their population changes; and about competition between species and predator/prey relationships. Nations generally do not provide adequate support for fishery research and development and government bodies have little power to manage fisheries and conserve resources.

If high yields of fish are to be sustained, then fishing must be limited — but how? Quotas would be one answer, but this raises many national and international problems. Regulations are difficult to enforce, and as the resource becomes smaller, the squeeze becomes tighter and there is a greater suspicion that 'others' are not abiding by the regulations. This results in a tendency to flout the laws.

Fishing the open sea presents further international regulation problems. Who is to utilise the oceanic waters? The open sea is at the mercy of anyone who has the technology to exploit it; who then is going to enforce protective regulations? It is possible that the Food and Agriculture Organization of the United Nations (FAO) will eventually decree that the open seas are internationally owned, so that the resources can be managed and conserved for the benefit of all nations.

Perhaps another way of harvesting food from the sea is by fishing the food of the fish — that is, the planktonic animals or 'krill'. But this food is not concentrated in one area like the fish, and the fishing effort required uses too much money to obtain a good yield. Although krill fishing has been carried out in certain parts of the world, much more work needs to be done before fishing for krill becomes a truly viable proposition. In addition, an intense krill fishery may affect the recovery of some of the large whales (page 131).

In the future, deep bottom-living fish may be exploited, but once again, much more research is required into fishing methods so that a good yield can be sustained while the resources are protected from over-utilisation.

Fish farming would seem an obvious way to revive the declining fish stocks. The idea is to raise young fish from the egg, through the earlier months, and then use them to restock the fishing grounds. But a large amount of capital is required, and it takes a long time before there is any return on the investment. There is also a danger of the fishing grounds being fished by unauthorised operators, and then the investment never returns to the restocking country. Sadly, one of the greatest threats to fish farming is from pollution because many farms are near to large industrial areas and coastal populations.

The fishing industry has many problems, but solutions will

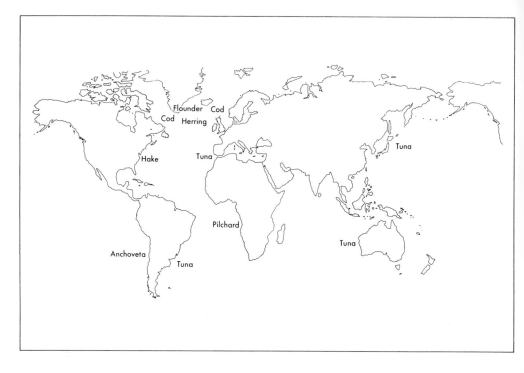

Figure 13.1: Fully exploited and over-exploited fisheries of the world

eventually result from increased research into the living resources of the sea. This will indicate what improvements and regulations are necessary to increase real fishing efficiency. We also need protective regulations for international waters where there is a question of who owns the resource. These areas should be available to all nations — and not just to those that have the necessary technology to exploit the open sea.

MINERAL RESOURCES

There are over 350 million cubic miles of ocean, and each cubic mile (almost five billion tons of water) contains about 165 million tons of solids (Table 13.1). Yet only salt, bromine, magnesium and fresh water are extracted from the sea commercially.

The sea-bed contains many minerals in scattered localised deposits lying on, or contained in, the sea floor. These include:

1. Fluid and dissolved minerals that are extracted from boreholes. (Oil and gas account for about 90 per cent of all minerals extracted from the sea, but sulphur and potash are also extracted through boreholes, using solution methods.)

2. Solid deposits such as coal, iron ore, nickel-copper ores, tin and limestone that are mined in tunnels originating on land.

3. Other surface deposits like shell, sand, gravel, diamonds and lumps of manganese and phosphorite that are dredged, though as yet the industry is confined to the continental shelf.

There are many obstructions to commercial exploitation of the seas, such as a lack of knowledge about the distribution of resources and about the possible damage to wildlife. In deeper waters there is also the question of ownership. Should the countries that are techno-

Table 13.1: The dissolved solids in seawater

SUBSTANCE	TONS PER CUBIC MILE	SUBSTANCE	TONS PER CUBIC MILE	SUBSTANCE	TONS PER CUBIC MILE
Chlorine	89,500,000	Indium	94	Silver	1
Sodium	49,500,000	Zinc	47	Lanthanum	1
Magnesium	6,400,000	Iron	47	Krypton	1
Sulphur	4,200,000	Aluminium	47	Neon	0.5
Calcium	1,900,000	Molybdenum	47	Cadmium	0.5
Potassium	1,800,000	Selenium	19	Tungsten	0.5
Bromine	306,000	Tin	14	Xenon	0.5
Carbon	132,000	Copper	14	Germanium	0.3
Strontium	38,000	Arsenic	14	Chromium	0.2
Boron	23,000	Uranium	14	Thorium	0.2
Silicon	14,000	Nickel	9	Scandium	0.2
Fluorine	6,100	Vanadium	9	Lead	0.1
Argon	2,800	Manganese	9	Mercury	0.1
Nitrogen	2,400	Titanium	5	Gallium	0.1
Lithium	800	Antimony	2	Bismuth	0.1
Rubidium	570	Cobalt	2	Niobium	0.05
Phosphorus	330	Cesium	2	Thallium	0.05
Iodine	280	Cerium	2	Helium	0.03
Barium	140	Yttrium	1	Gold	0.02

Source: Wenk, E. jr. The Physical Resources of the Ocean. In: Ocean Science
Scientific American 1977. Pub: Freeman and Co., San Francisco.

logically advanced be allowed to exploit international waters? And what about a share for the less developed nations? As technologically advanced nations become rapidly more depleted in resources this conflict of ownership will become more important. As with the fishing industry perhaps we should adopt international ownership of the waters so that they can be mined for the benefit of all.

Far more knowledge is required about the sea floor, its minerals, possible ways of mining the resources, and about the effects this will have on the environment. With this knowledge we must also make regulations to ensure the protection of the wildlife and the resources, and to ensure that the proceeds of exploitation are distributed fairly.

TRANSPORT

The oceans have given us the means to spread all over the world. We can also send goods across the seas, and today practically every nation has its fleet of merchant vessels. In the days of sail, before synthetic products resistant to natural breakdown were invented, there was little pollution to worry about. Today there are thousands of large ships, which carry thousands of tons of fuel; there are tankers that carry two or three hundred thousand tons of crude oil for refining; and there are ships that carry many other dangerous cargoes — poisonous materials, refinery products and gases. Each ship has a coating of poisonous antifouling paint that slowly dissolves into the sea. The people on board use food and stores that are wrapped in plastic, glass and tin, most of which is disposed of at sea.

Oil pollution entered the public eye in 1967 with the *Torrey Canyon* disaster. It has stayed at the forefront of debate ever since, despite the fact that of all the oil input into the sea — about 6 million metric tons per year — just over four per cent (300,000 tonnes) is due to shipping accidents. About 10 per cent of the oil input is due to natural seepage, which means that almost 86 per cent of the oil entering the sea is from urban and industrial developments and from shipping operations (Table 13.2).

Table 13.2: The estimated amount of petroleum hydrocarbon input to the sea (in metric tons per annum)

SOURCE	AMOUNT	PERCENT
Natural seepage	600,000	9.8
Atmospheric rainout	600,000	9.8
Offshore production	80,000	1.3
Coastal refineries	200,000	3.3
Coastal municipal	300,000	4.9
Coastal non-refining	300,000	4.9
Urban run-off	300,000	4.9
River run-off	1,600,000	26.2
Total from non-shipping	3,980,000	65
Tankers LoT*	310,000	5.0
Tankers non-LoT*	770,000	12.6
Drydocking and repair	250,000	4.1
Terminal operations	3,000	0.05
Bilges and bunkering	500,000	8.2
Tanker accidents	200,000	3.3
Non-tanker accidents	100,000	1.6
Total from shipping	2,133,000	35
Total input to sea	6,113,000 per annum	

*LoT — indicates Load on Top — the new cargo is loaded on top of the old washings.

Source: United States Ocean Affairs Board 1975. Cited by
A.D. Couper and W. Burger: Analysis of Shipping Casualties in the Region. In: *The Protection of the Coast of Wales from Pollution from Ships and Terminal Sources*. 19 September 1979. Nautical Institute Publication.

The chances of accidents should be diminished as far as possible, but much could be done to decrease the oil input to the sea if more attention was paid to these other causes.

Garbage pollution of the sea has, over the last few years, shown itself to be a significant problem. Fishermen report finding plastic cups in the stomachs of fish; young seabirds are strangled in their nests because pieces of synthetic fibre have been used for nesting material; turtles have suffered from eating plastic bags in mistake for jellyfish, and from attempting to swim while wrapped in plastic sheeting.

In 1973, the International Maritime Organisation (IMO — formerly IMCO) produced regulations for preventing the disposal of garbage at sea. Twelve years later the member governments have still not adopted the regulations. Shipping companies, owners and suppliers should ensure that vessels are provided with adequate disposal facilities, and that ships do not take on board large amounts of non-biodegradable material. Governments should also provide every port with adequate facilities for the disposal of garbage.

RECREATION

The sea offers facilities for many activities: boating, swimming, fishing, diving, water-skiing, walking along the coastline or simply sitting on a beach in the sun. Each of these activities must be limited and controlled to ensure the enjoyment of all parties (Colour Plate 64). It should also be possible to enjoy the sea without the risk of being poisoned in the water, getting oil on skin and clothing, and having garbage on the beach which is unsightly and results in injuries.

It has been estimated that by the year 2000, over half the population of the United States will live on five per cent of the land in the megalopolises of the Atlantic and Pacific coasts and on the Great Lakes. This will make enormous demands on the coastline and on the ability of the sea to absorb waste.

DISPOSAL OF WASTE

The sea is the ultimate sink, receiving industrial and urban waste, chemicals from rain and rivers, and waste from ships. Man has reached the disposable era. Everything is wrapped in plastic; glass and tin are thrown away as if there was an infinite supply. It is important to find out just how much the sea can tolerate before this abuse becomes detrimental to man, wildlife and the environment (Table 13.3).

Table 13.3: Some marine pollution figures

Table 13.3: Some marine pollution figures

a) The top six groups of substances currently discharged in the sea which could be harmful* in sufficent quantities

1. Domestic sewage including food processing waste
2. Pesticides (including fungicides)
3. Inorganic wastes including heavy metals
4. Radioactive materials
5. Oil and oil dispersants
6. Petrochemicals and organic chemicals

* Harmful means harm to living resources, hazard to human health, hindrance to marine activities, and reduction of amenities according to a Group of Experts on the Scientific Aspects of Marine Pollution (a United Nations working group).

b) The types and amounts of material dumped into the oceans by the United States (in tons)

WASTE TYPE†	ATLANTIC	GULF	PACIFIC	TOTAL
Dredge spoils	15,808,000	15,300,000	7,320,000	38,428,000
Industrial wastes	3,013,200	696,000	981,300	
1. Acids				2,720,500
2. Refinery				562,900
3. Pesticides				328,300
4. Paper mill				140,700
5. Other				938,100
Sewage sludge	4,477,000	0	0	4,477,000
Construction debris	574,000	0	0	574,000
Solid waste	0	0	26,000	26,000
Explosives	15,200	0	0	15,200
Total	23,887,400	15,996,000	8,327,300	48,210,700

† No figures for the disposal of radioactive substances are available although they are dumped off all three coasts.

c) Estimates of deposition rates from the atmosphere over the North Sea (in metric tons per year)

SUBSTANCE	TOTAL
Dissolved organic matter	5,000,000
‡ Sulphur as sulphur dioxide	3,000,000
‡ Particles	1,500,000
Nitrogen (fixed)	1,000,000
Lead	15
DDt residues	30
Mercury	30

‡ An additional 1 ton of dust per square kilometre a year on average is likely to be deposited during 15 years in one century due to volcanic eruptions.

Sources: (a) J. Wardley Smith, 'Oil Exploration in the Sea — A Review of the Pollution Risk', in M.M. Sibthorp (ed.), Oceanic Management: Conflicting Uses of the Celtic Sea and Other Western UK Waters (Europa, 1977).
(b) Skinner, B.J. and K.K. Turekian, Man and Ocean. Foundations of Earth Science Series 1973. Figures from Ocean Dumping, 1970, Council on Environmental Quality.
(c) Sibthorp, M.M. (ed.) The North Sea: Challenge and Opportunity. Europa 1975. For the David Davies Memorial Institute of International Studies.

There are many difficulties in trying to measure the effect of waste disposal because the ocean is not the same everywhere: harbours, bays, estuaries and the open sea have different qualities that require different regulations. Neither is the sea itself pure, and there are many changing factors such as currents and waves that may spread substances to widely separated area. Against this ever-changing background the environmentalist must be able to assess the effects of discharges into the sea.

We need research into the natural changes in the oceanic environment because our knowledge is greatly lacking. With this information each input to the sea can be judged in the light of knowledge about the material and its effects. Some substances are directly harmful to marine life, and therefore also to man, and should be prevented from entering the sea at all. This category would include all radioactive material and synthetic organic compounds, like some pesticides, because marine life has no natural defences against these. Other substances are harmful above certain levels — such as mercury, chromium, lead, zinc, nickel and silver. More knowledge is required about what happens to these chemicals in nature, and what their effects are: some organisms are more sensitive than others. Industries should decrease their output of these chemicals to the lowest level technologically possible.

Municipal sewage waste can be a health hazard and may be aesthetically displeasing, but the sea can absorb a large amount of sewage if it is broken up and dispersed over a wide area. This requires a great deal of information about the disposal site. Will the currents and tide bring 'floatables' back to the shore? Will the sewage disperse or collect in one place? What amount of sewage can the area tolerate?

Rivers and rain are constantly carrying pollutants into the sea — notably pesticides, exhaust chemicals from cars and industry, and petroleum products. It is also vital to anticipate pollutants that may become serious as new compounds are produced. We need more information about the natural environment and a consideration of which materials can be disposed of at sea without adverse effects. Such knowledge will lead to a large saving of resources and an unpolluted ocean.

THE NEED FOR CO-OPERATION

One nation alone trying to reconcile the conflicting uses of the sea would have difficulties. When the interests of many nations are added the possibilities of co-operation seem remote. But some international organisations have developed with some degrees of success. The North East Atlantic Fisheries Commission arose in response to increasing pressure on fisheries, and there are international waste and pollution regulations. But much more remains to be done.

Great problems have arisen through the allocation of oceanic resources. The technologically advanced nations bordering the sea argue that they have the right to the resources within 'their' waters. In 1958 the country could continuously develop the resources for as deep as they could go. This supported President Truman's declaration that the US owned the minerals on its continental shelf.

But with improved technology more resources are within reach and problems have arisen to do with the question of ownership — freedom of the high seas no longer was fair.

It seems that there are several types of countries: (1) those with the technology argue that resources should be available to those with the ability, i.e. themselves; (2) countries without the technology (mostly underdeveloped) but with certain resources that they export would prefer minimal exploitation of some marine resources; (3) the underdeveloped countries that have neither the technology nor the material to export, and purely want some benefit from the exploitation of 'international' waters.

The problems with territorial waters have been largely the concern of the United Nations Conference on the Law of the Sea which has put forward the idea of an economic exclusion zone of 200 miles within which states have the right to the resources. This may go some way to helping, but underdeveloped countries will now have ownership over areas that they still cannot exploit. They must either wait until they develop the technology or must enter into bilateral agreements with developed nations. The agreement is usually to aid the poorer country in return for being allowed to, for example, fish within their economic exclusion zone. These agreements sound like a good answer but there are problems with just how much 'aid' is given and how useful it is — but this takes the arguments beyond the scope of this discussion. However, it is worth noting that some of the poorest areas of the world lie next to some of the richest sea areas!

THE PRESERVATION OF MARINE LIFE

Many, varied and conflicting are the uses of the sea. No one would dispute the importance of preserving marine life, because the quality and continuation of man's existence on earth is determined by the maintenance of an unpolluted ocean. But the sea provides a vast store of resources which, if used wisely, can be of immense benefit to man. But single-purpose uses motivated by short-term advantages for individuals, industry or governments will lead to gross exploitation and ultimately destruction. There must be far less competition and more co-operation to ensure that everyone has a share. This will lead to unpolluted oceans, a conservation of resources and — much needed today — national and international co-operation to ensure the survival of the sea and, ultimately, of humankind.

Maps

1. Areas of high productivity

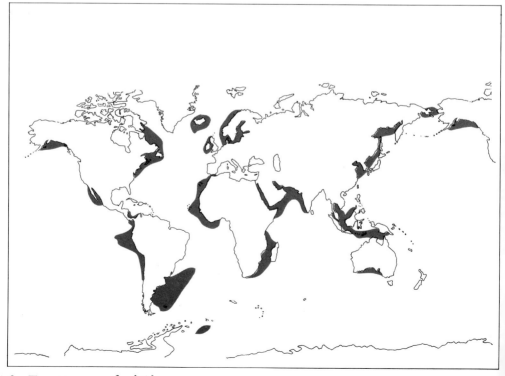

2. The occurrence of red tides

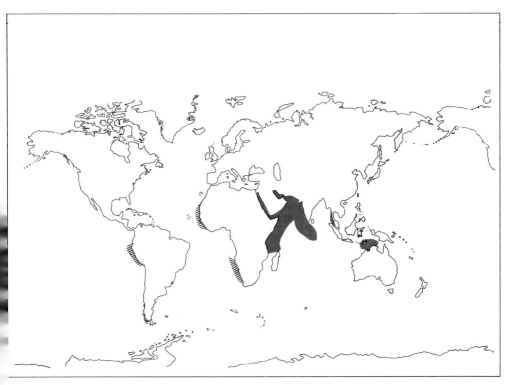

3. Areas where milky seas have been recorded

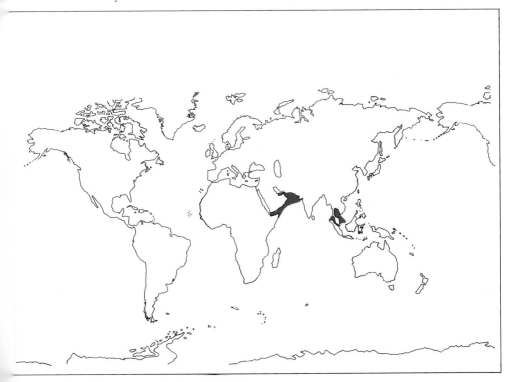

4. Areas where luminescent wheels are found

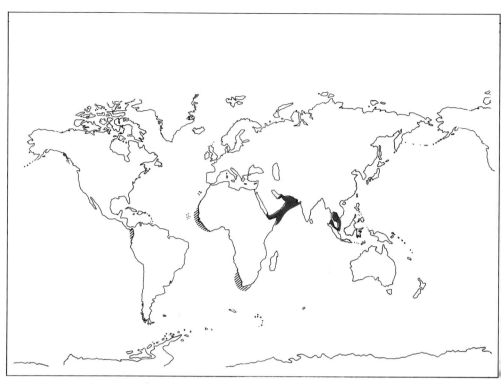

5. Areas where luminescent bands are found

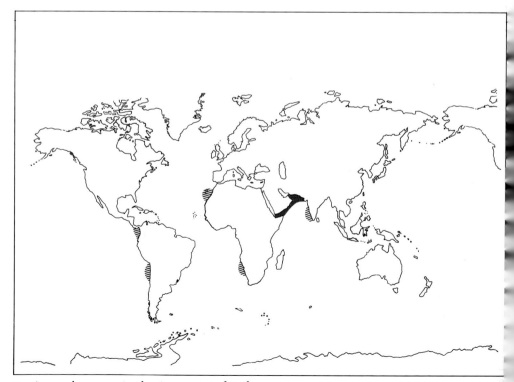

6. Areas where erupting luminescence is found

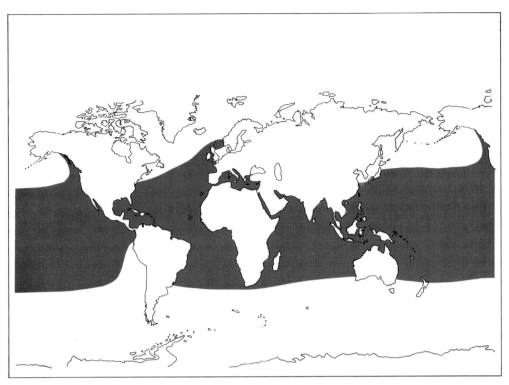

7. Distribution of the by-the-wind.sailor, *Velella velella*

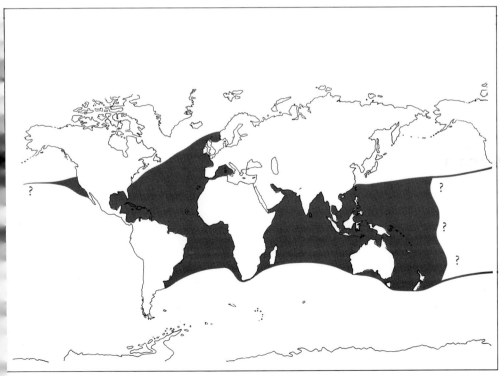

8. Distribution of the Portuguese man-of-war, *Physalia physalis*

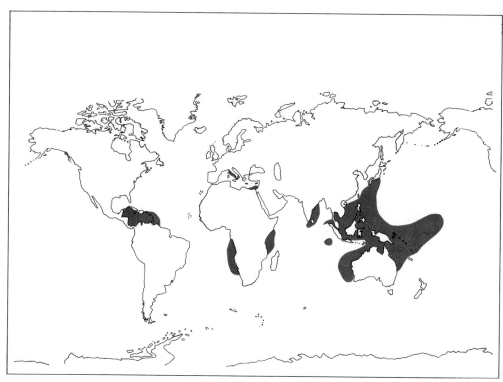

9. Distribution of the box jellyfish, Cubozoa

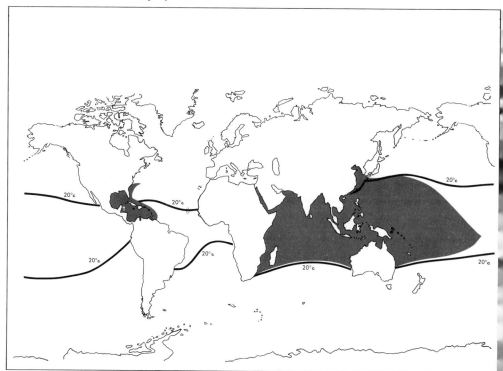

10. The coral areas of the world

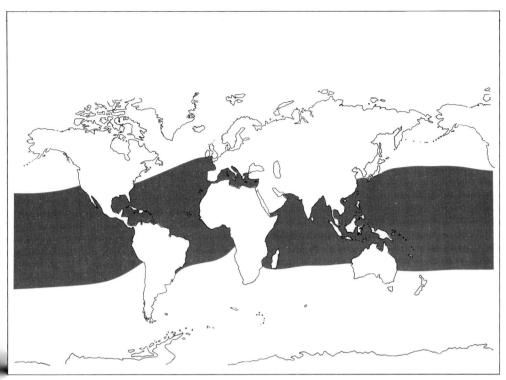

11. Distribution of the purple bubble-raft snail, *Janthina* spp.

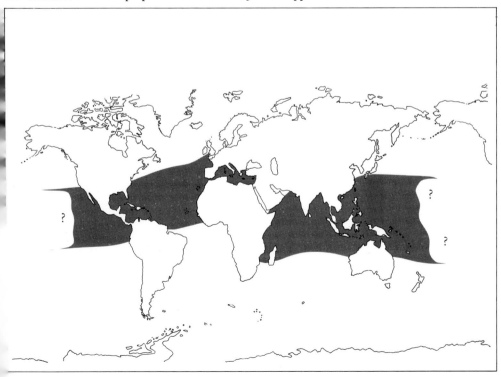

2. Distribution of the planktonic sea slug, *Phylliroe bucephala*

171

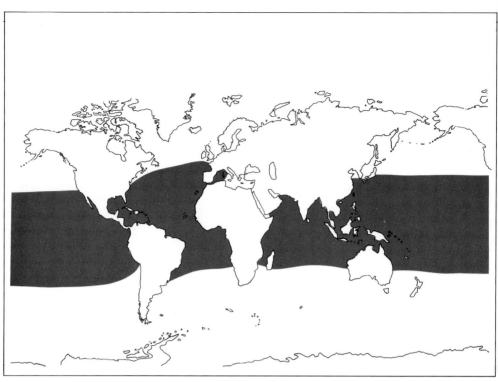

13. Distribution of the planktonic sea slug, *Glaucus atlanticus*

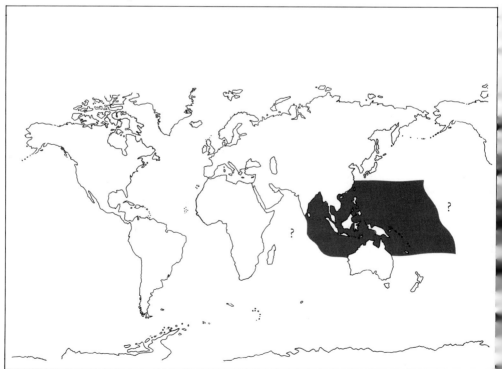

14. Distribution of *Nautilus*

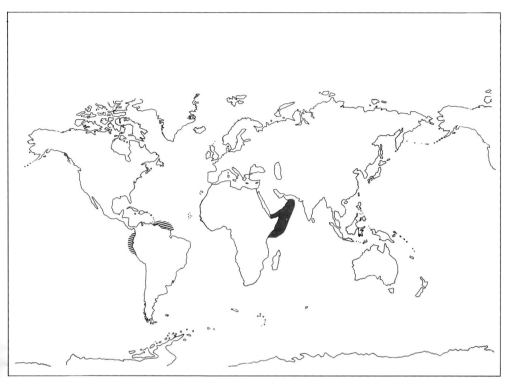

15. Distribution of the swimming crab, *Charydris* spp.

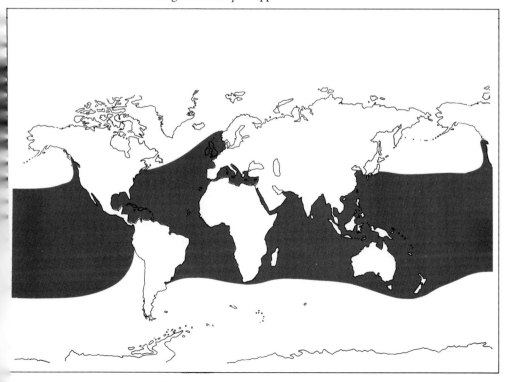

16. Distribution of the colonial sea squirt, *Pyrosoma* spp.

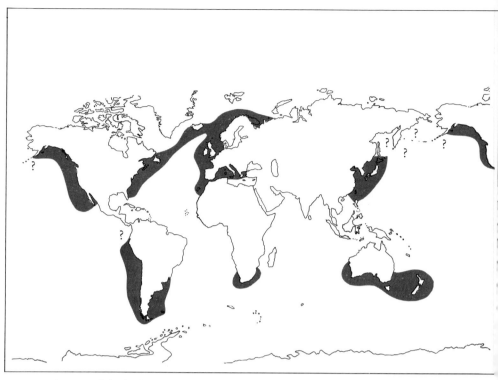

17. Distribution of the basking shark, *Cetorhinus maximus*

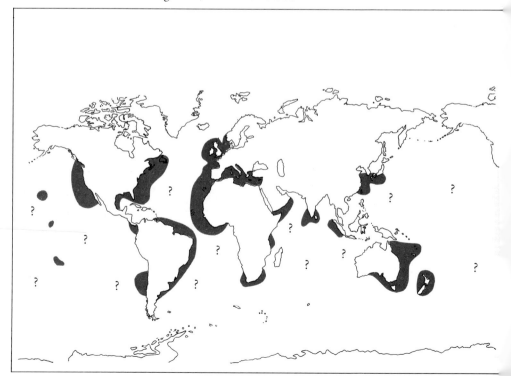

18. Distribution of the thresher shark, *Alopias vulpinus*

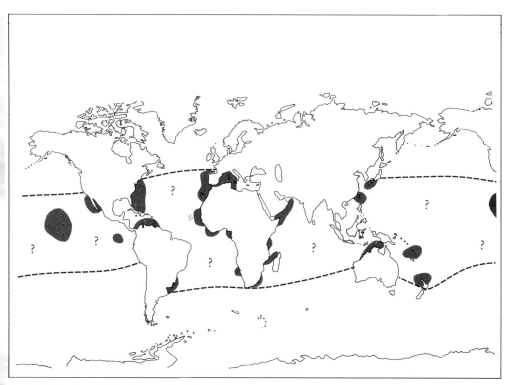

19.   Distribution of the big-eye thresher, *Alopias superciliosus*

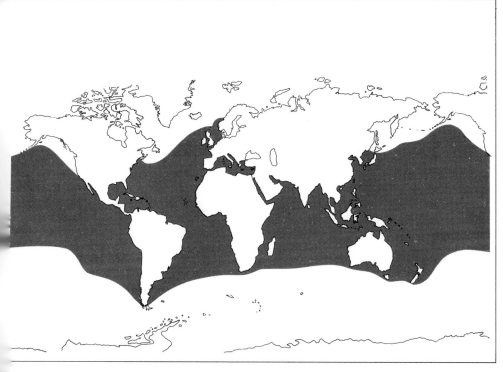

20.   Distribution of the mako, *Isurus oxyrinchus*

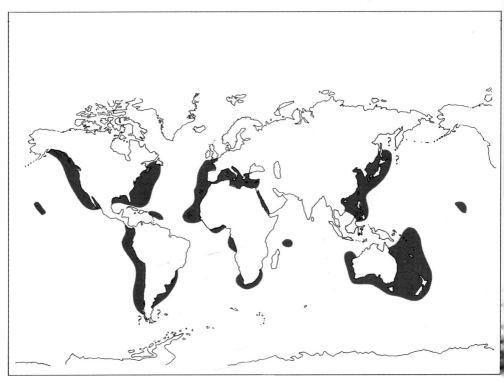

21. Distribution of the great white shark, *Carcharadon carcharias*

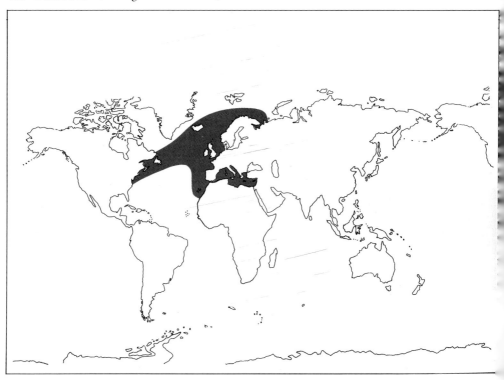

22. Distribution of the porbeagle, *Lamna nasus*

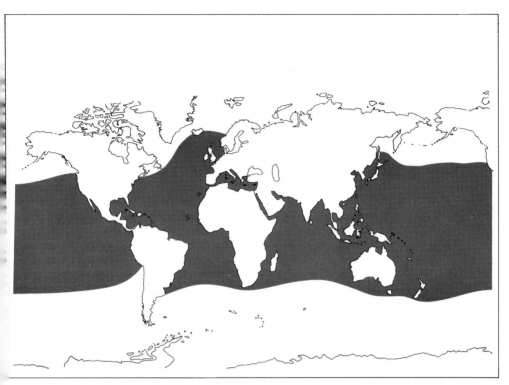

23. Distribution of the tiger shark, *Galeocerdo cuvier*

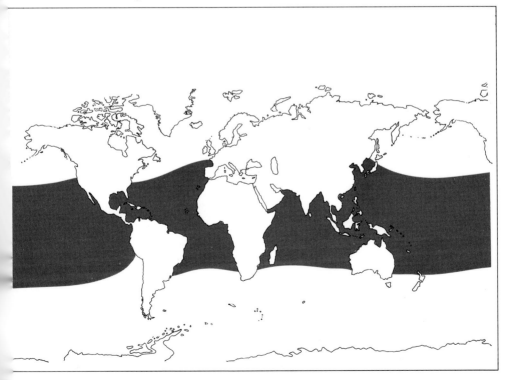

4. Distribution of the white tip shark, *Carcharinus longimanus*

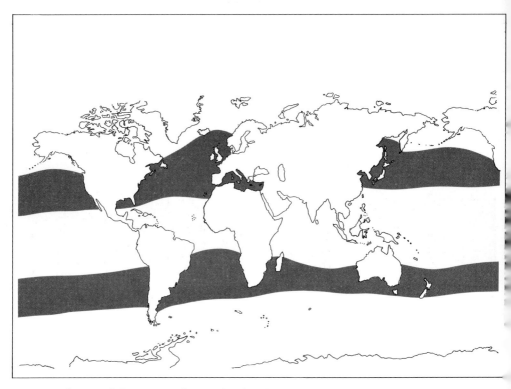

25. Distribution of the common hammerhead, *Sphyrna zygaena*

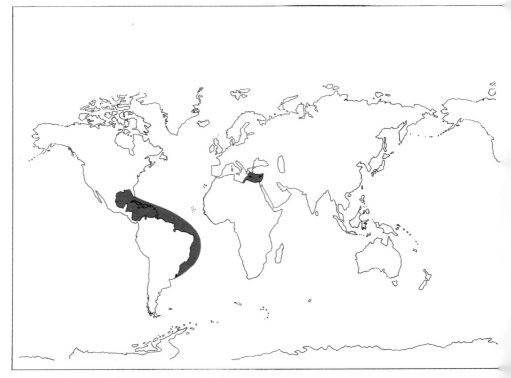

26. Distribution of the great hammerhead, *Sphyrna tudes*

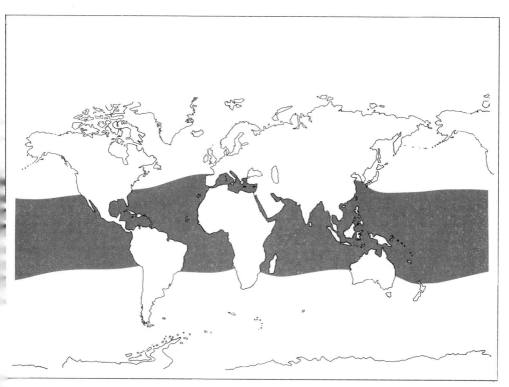

27. Distribution of the scalloped hammerhead, *Sphyrna lewini*

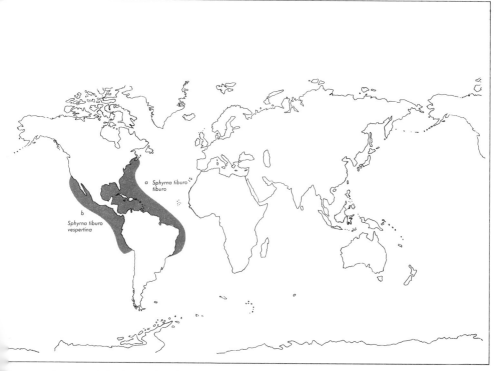

8. Distribution of the bonnethead, *Sphyrna tiburo*

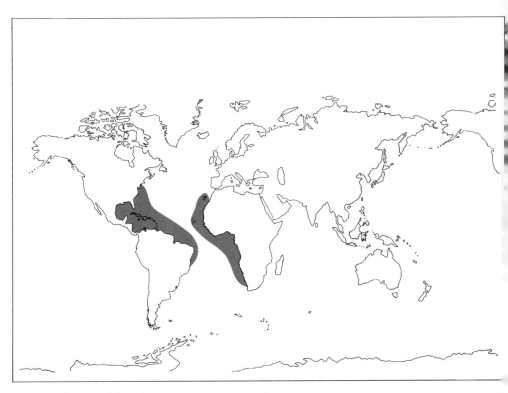

29. Distribution of the Atlantic manta ray, *Manta birostris*

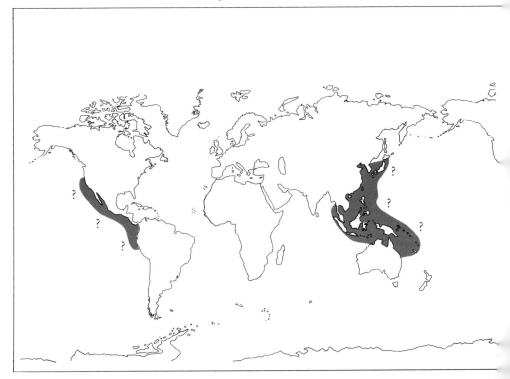

30. Distribution of the Pacific manta ray, *Manta hamiltoni*

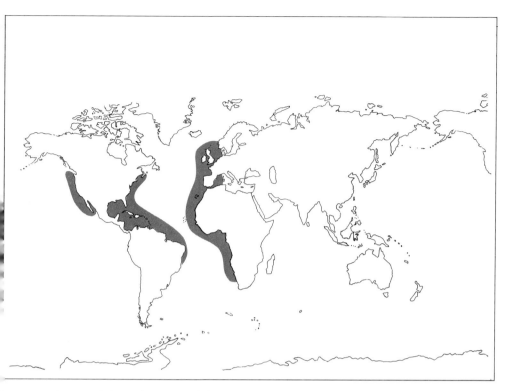

31. Distribution of the eagle rays, *Myliobatis* spp.

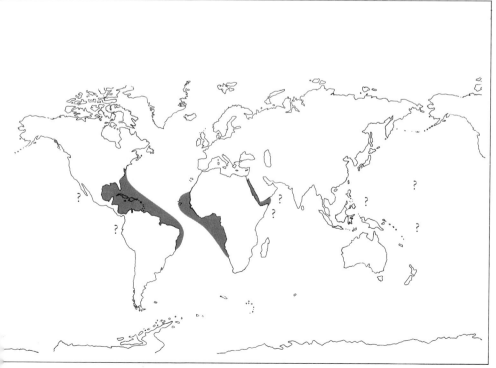

2. Distribution of the spotted eagle rays, *Aetobatus* spp.

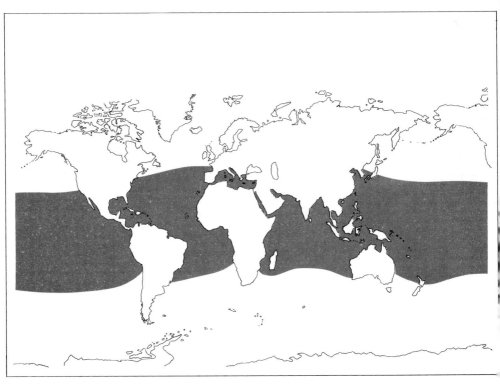

33. Distribution of the flying fish, Exocoetidae

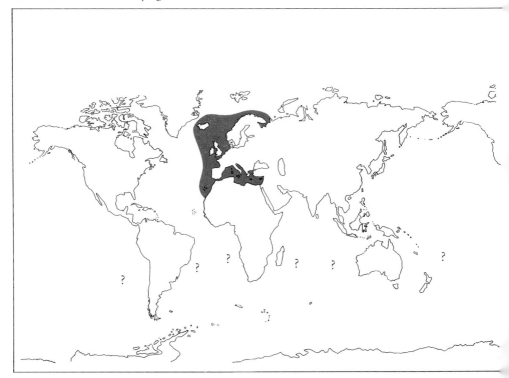

34. Distribution of the Atlantic skipper, *Scomberesox saurus*

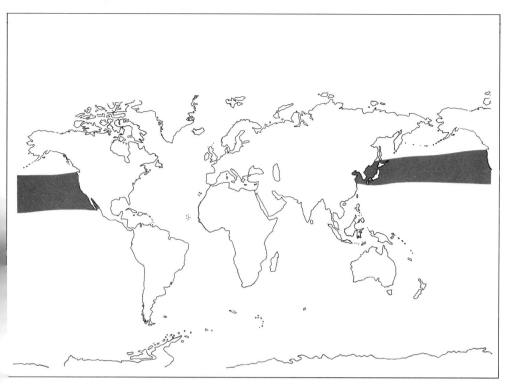

35. Distribution of the Pacific skipper, *Cololabis saira*

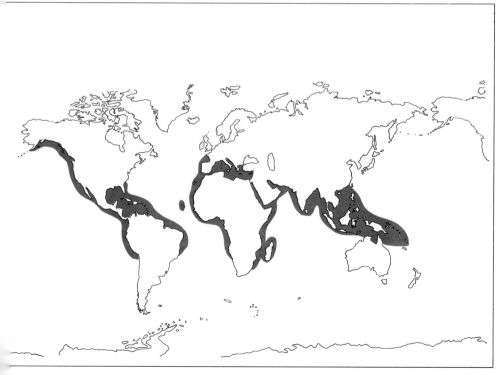

6. Distribution of the barracuda, Sphyraenidae

183

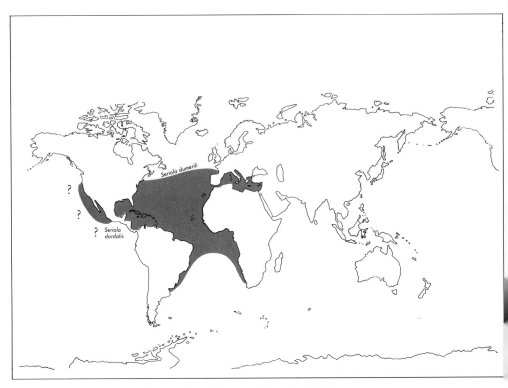

37. Distribution of the Atlantic and California yellowtails, *Seriola* spp.

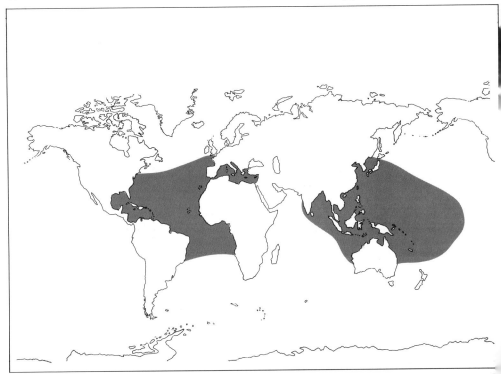

38. Distribution of the dolphin fish, *Coryphaena* spp.

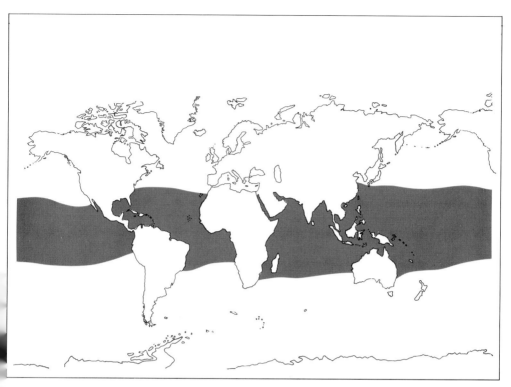

39. Distribution of the sailfish, *Istiophorus* spp.

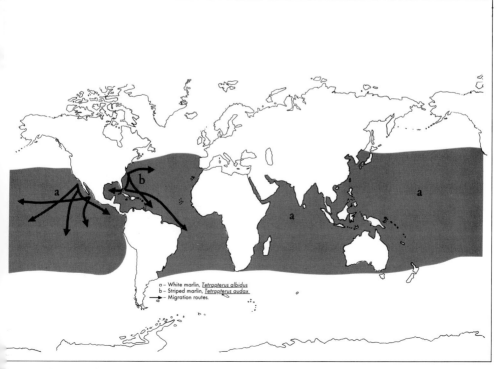

a – White marlin, *Tetrapterus albidus*
b – Striped marlin, *Tetrapterus audax*
→ – Migration routes.

). Distribution of the marlins, *Tetrapterus* spp.

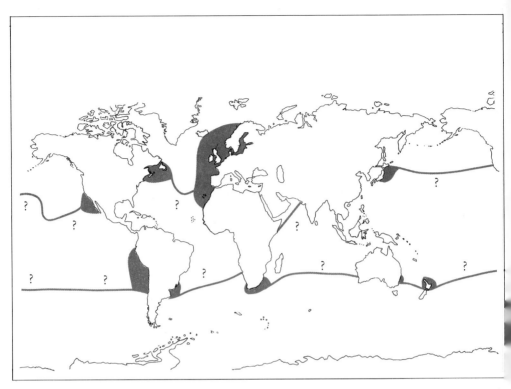

41. Distribution of the swordfish, *Xiphias gladius*

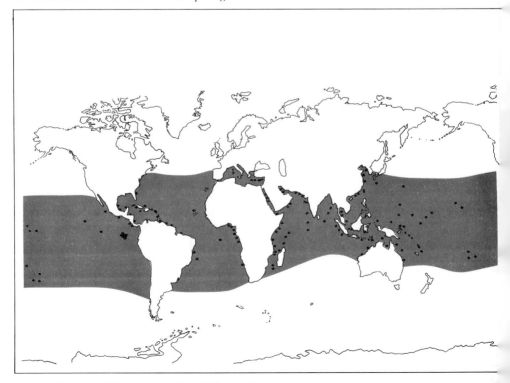

42. Distribution of the green turtle, *Chelonia mydas*

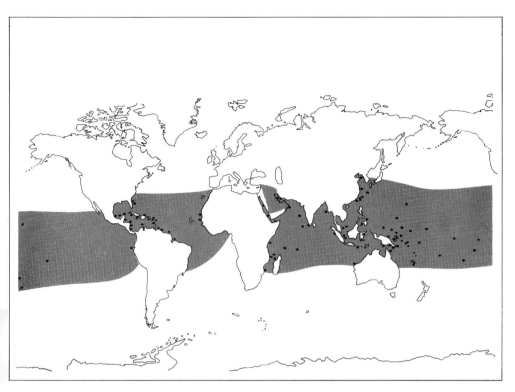

43. Distribution of the hawksbill turtle, *Eretmochelys imbricata*

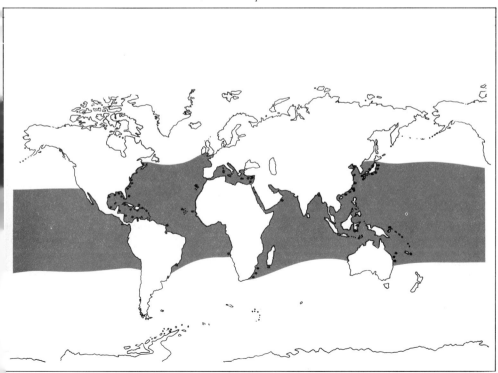

44. Distribution of the loggerhead turtle, *Caretta caretta*

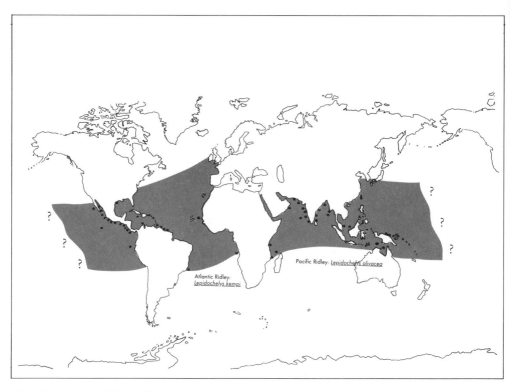

45. Distribution of the Ridley turtles, *Lepidochelys* spp.

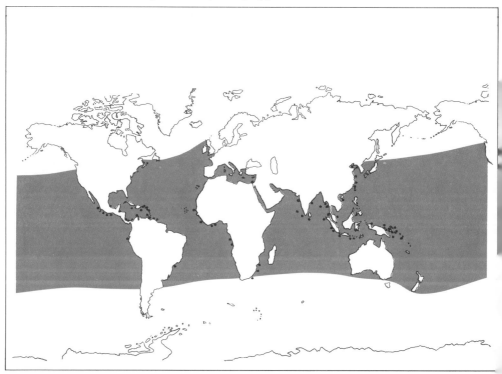

46. Distribution of the leatherback turtles, *Dermochelys coriacea*

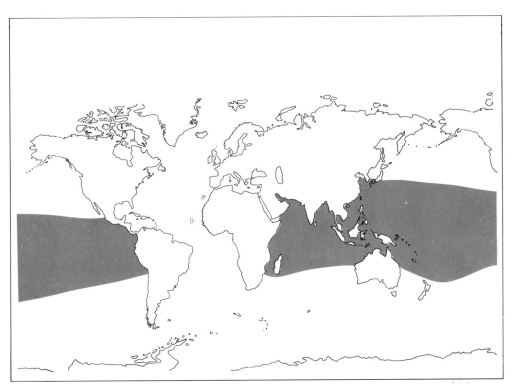

47. Distribution of sea snakes, Hydrophiidae

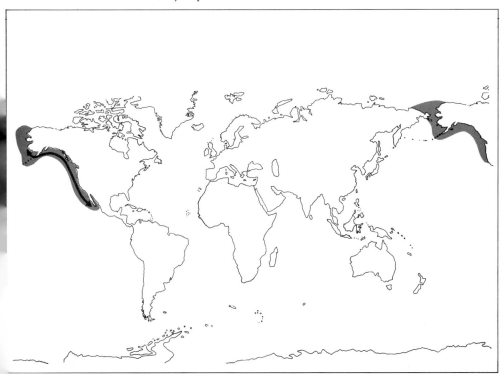

48. Distribution of the California grey whale, *Eschrichtius robustus*

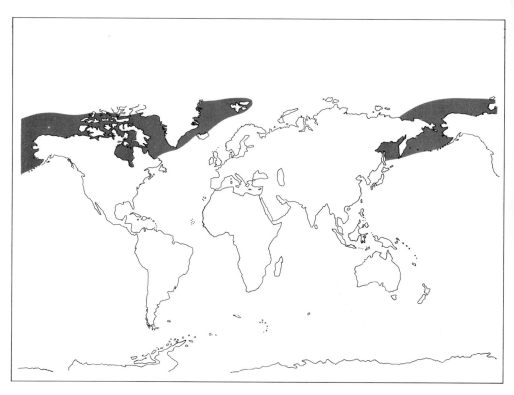

49. Distribution of the Greenland right whale, *Balaena mysticetus*

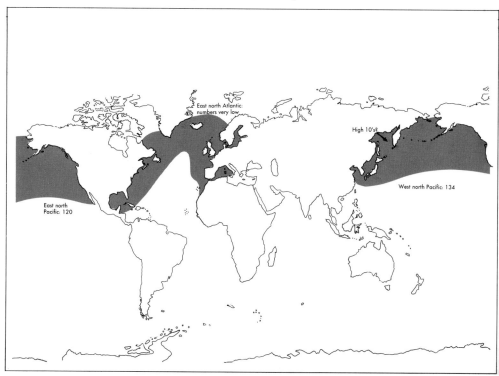

50. Distribution of the northern right whale, *Balaena glacialis*

190

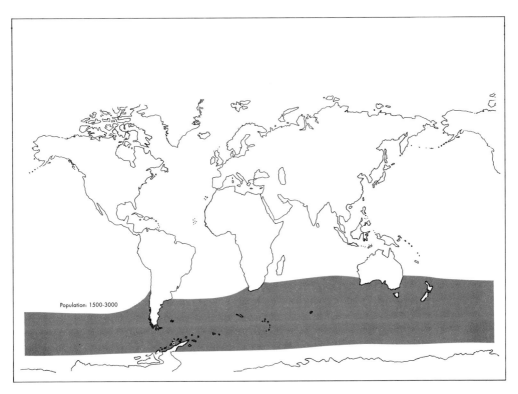

Population: 1500-3000

51. Distribution of the southern right whale, *Balaena glacialis australis*

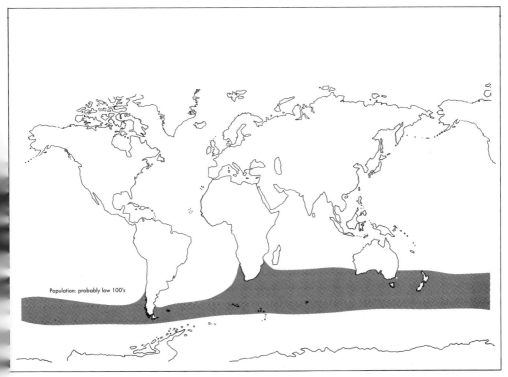

Population: probably low 100's

52. Distribution of the pygmy right whale, *Caperea marginata*

191

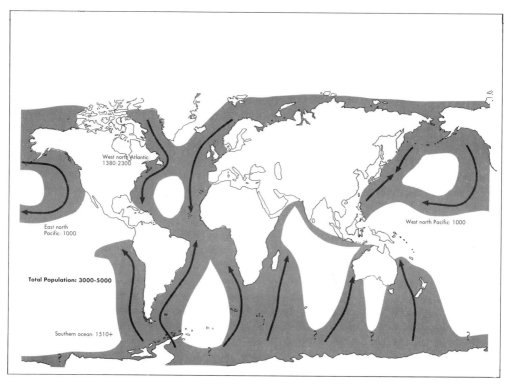

53. Distribution of the humpback whale, *Megaptera novaeangliae*

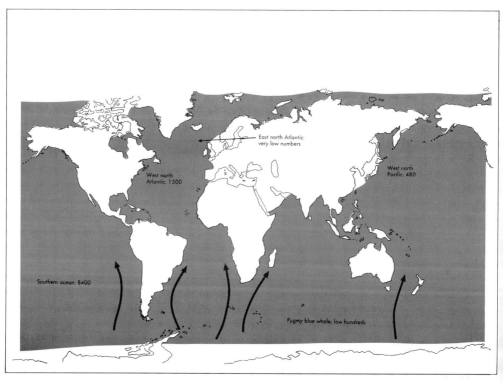

54. Distribution of the blue whales, *Balaenoptera musculus* and *Balaenoptera musculus brevicauda* — the pygmy blue

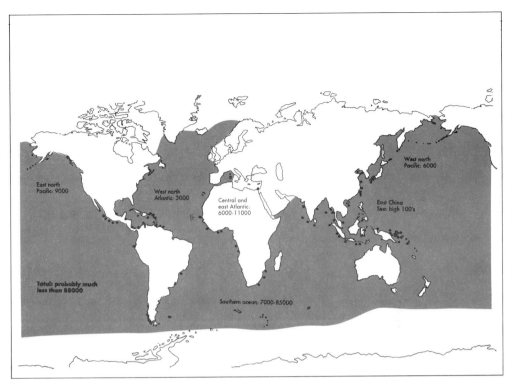

55. Distribution of the fin whale, *Balaenoptera physalus*

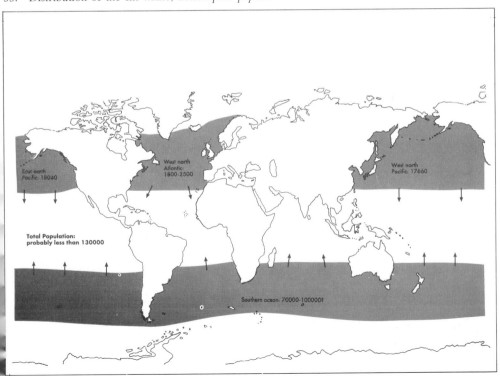

56. Distribution of the sei whale, *Balaenoptera borealis*

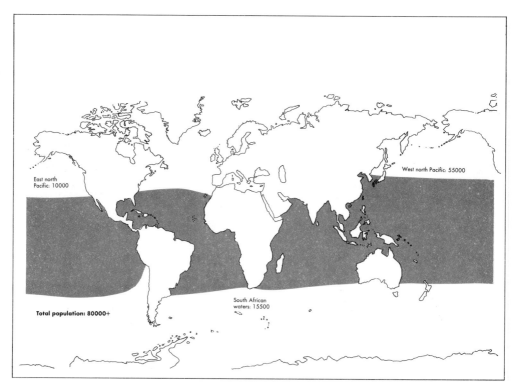

57. Distribution of Bryde's whale, *Balaenoptera edeni*

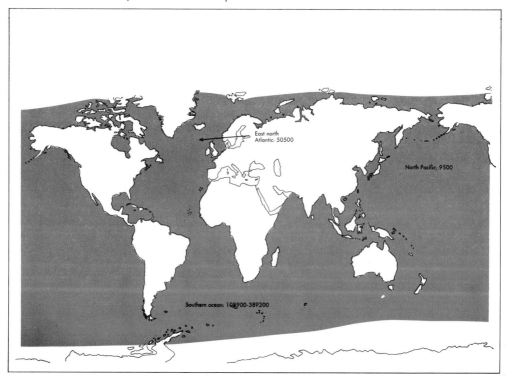

58. Distribution of the minke whale, *Balaenoptera acutorostrata*

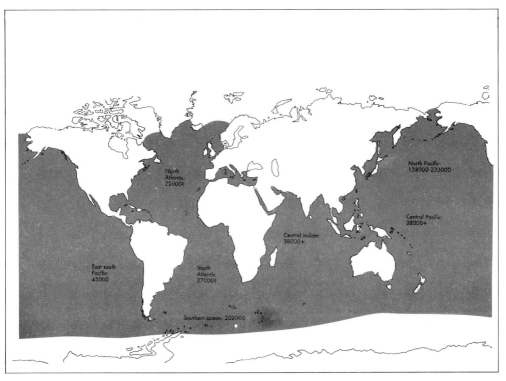

59. Distribution of the sperm whale, *Physeter macrocephalus*

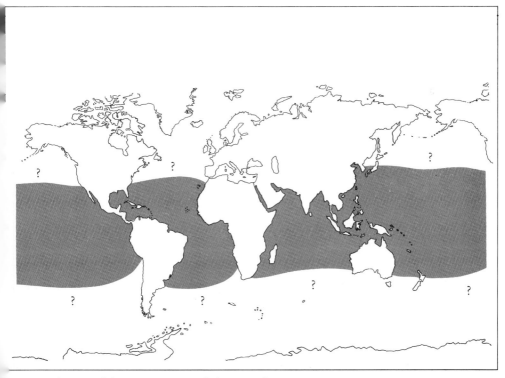

60. Distribution of the pygmy sperm whale, *Kogia breviceps*

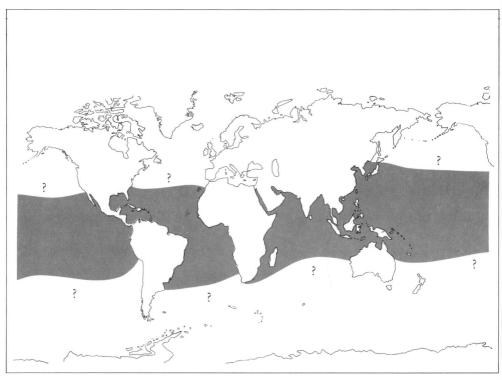

61. Distribution of the dwarf sperm whale, *Kogia simus*

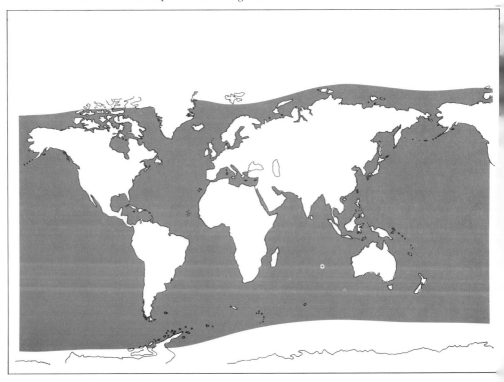

62. Distribution of the killer whale, *Orcinus orca*

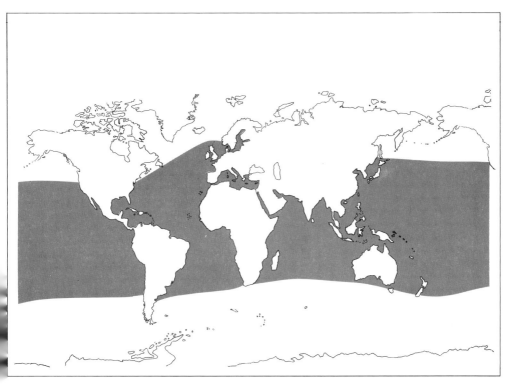

63. Distribution of the false killer whale, *Pseudorca crassidens*

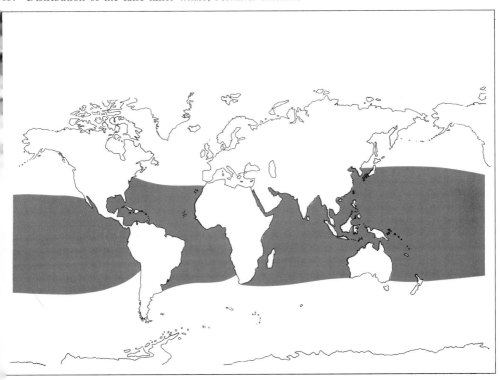

64. Distribution of the pygmy killer whale, *Feresa attenuata*

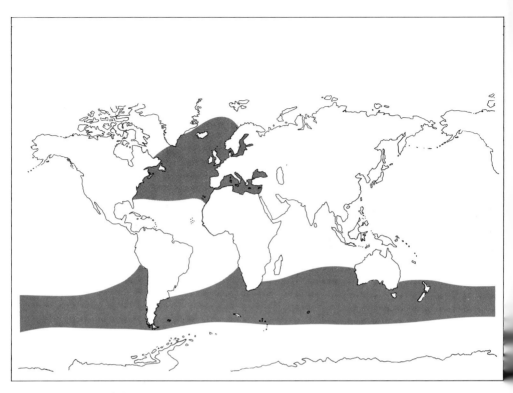

65. Distribution of the longfin pilot whale, *Globicephala malaena*

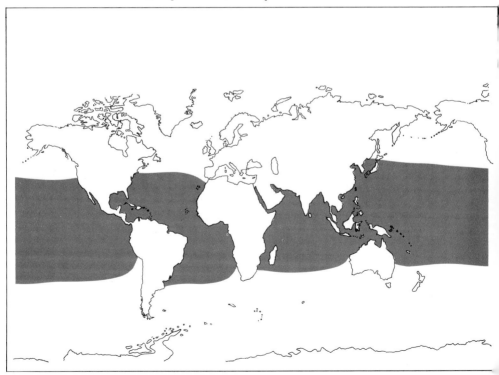

66. Distribution of the shortfin pilot whale, *Globicephala macrorhynchus*

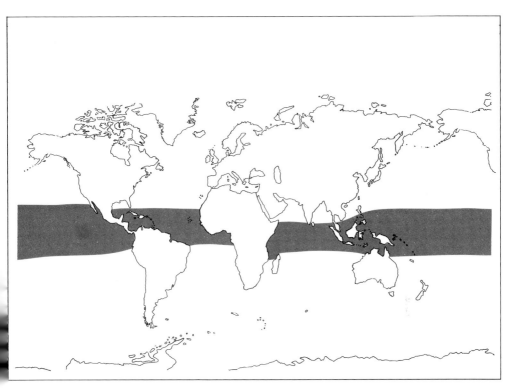

67. Distribution of the melonheaded whale, *Peponocephala electra*

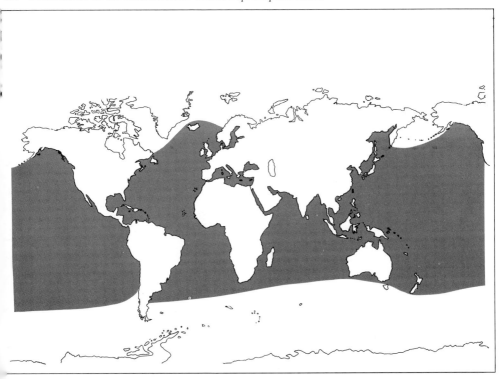

68. Distribution of the bottlenose dolphin, *Tursiops truncatus*

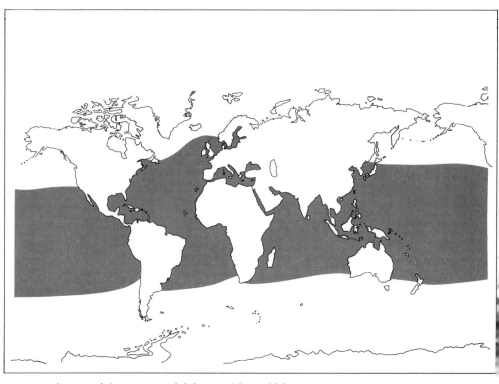

69. Distribution of the common dolphin, *Delphinus delphis*

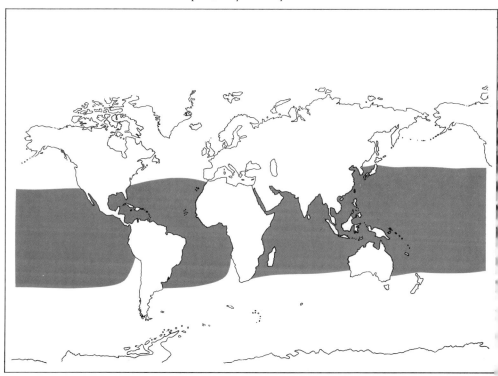

70. Distribution of the spinner dolphin, *Stenella longirostris*

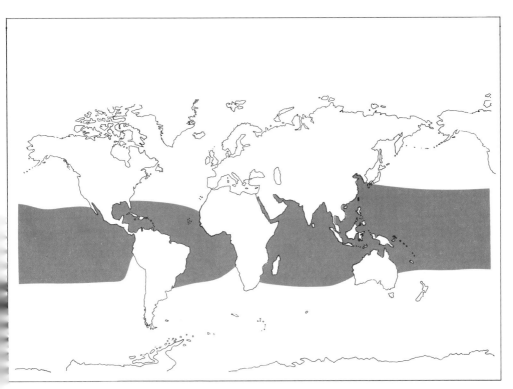

71. Distribution of the bridled dolphin, *Stenella attenuata*

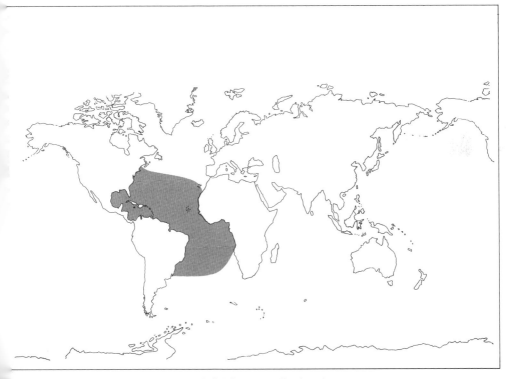

72. Distribution of the Atlantic spotted dolphin, *Stenella plagiodon*

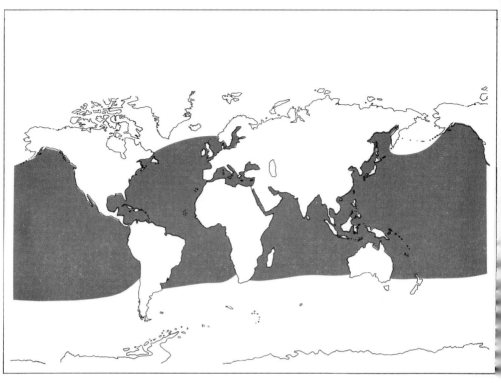

73. Distribution of the striped dolphin, *Stenella coeruleoalba*

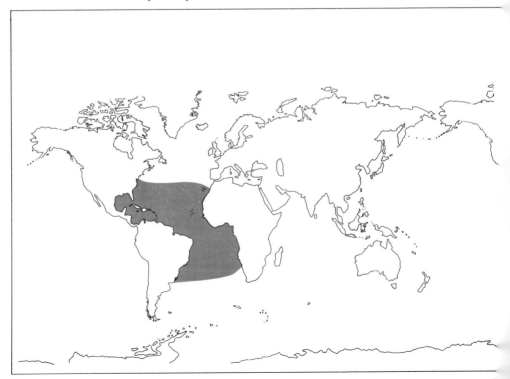

74. Distribution of the clymene dolphin, *Stenella clymene*

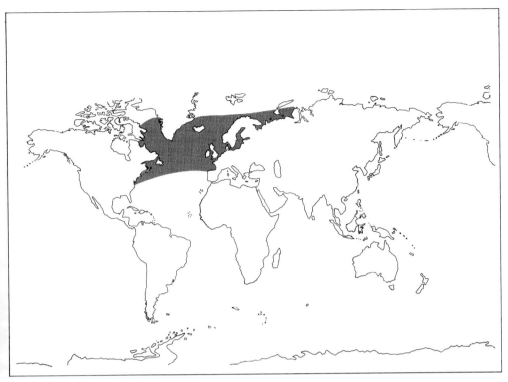

75. Distribution of the whitebeak dolphin, *Lagenorhynchus albirostris*

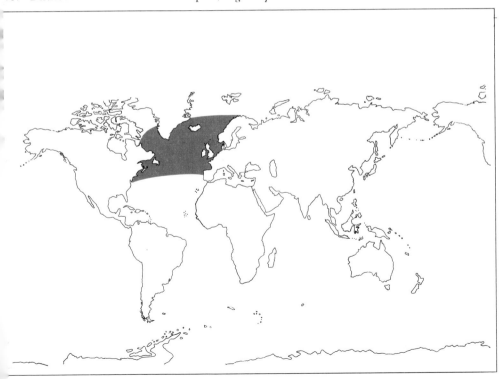

76. Distribution of the Atlantic whitesided dolphin, *Lagenorhynchus acutus*

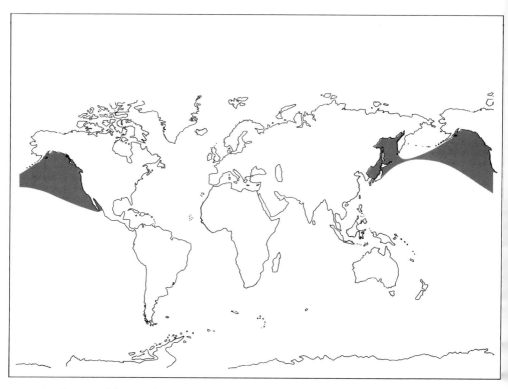

77. Distribution of the Pacific whitesided dolphin, *Lagenorhynchus obliquidens*

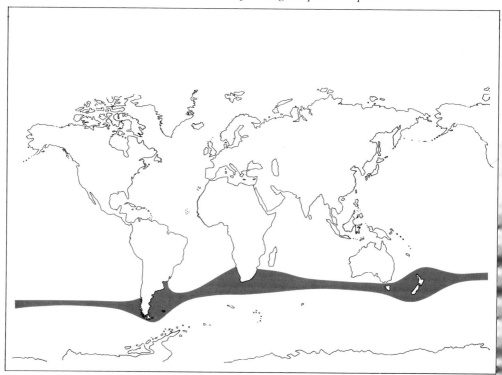

78. Distribution of the dusky dolphin, *Lagenorhynchus obscurus*

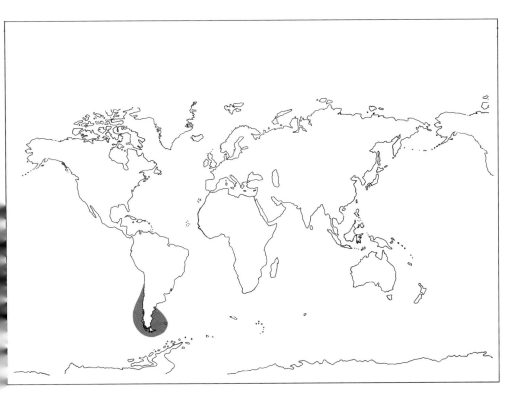

79. Distribution of Peale's dolphin, *Lagenorhynchus australis*

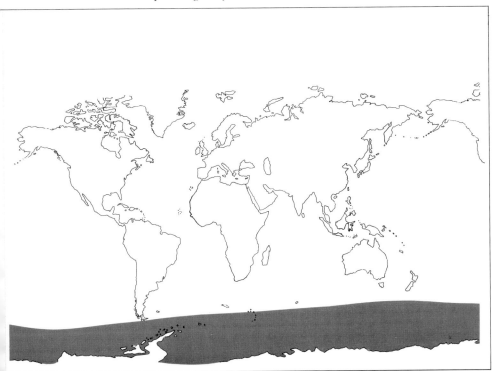

80. Distribution of the cruciger dolphin, *Lagenorhynchus cruciger*

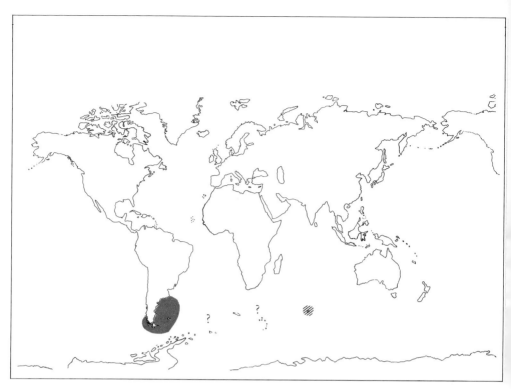

81. Distribution of the Commerson's dolphin, *Cephalorhynchus commersoni*

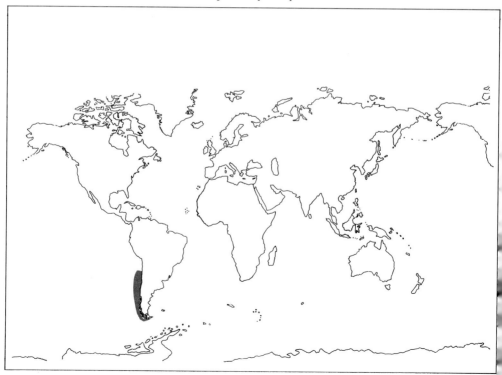

82. Distribution of the Chilean or black dolphin, *Cephalorhynchus eutropia*

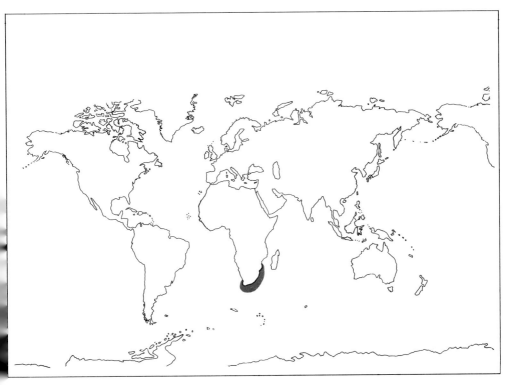

83. Distribution of the Heaviside's dolphin, *Cephalorynchus heavisidi*

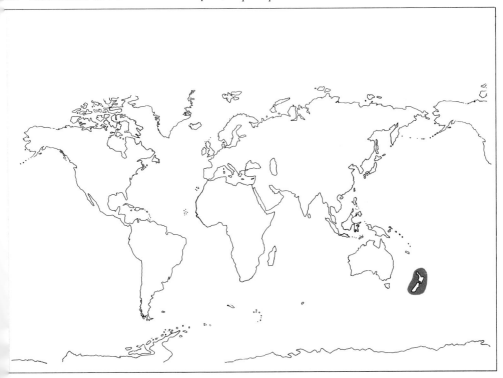

84. Distribution of Hector's dolphin, *Cephalorynchus hectori*

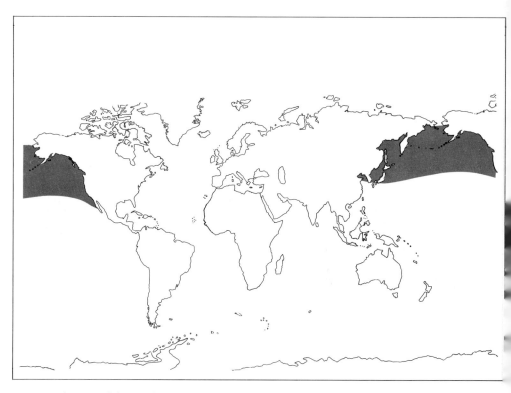

85. Distribution of the northern right whale dolphin, *Lissodelphis borealis*

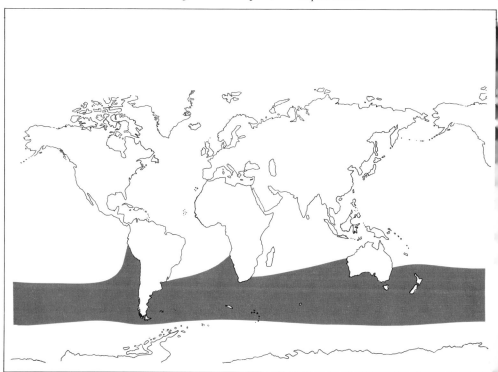

86. Distribution of the southern right whale dolphin, *Lissodelphis peroni*

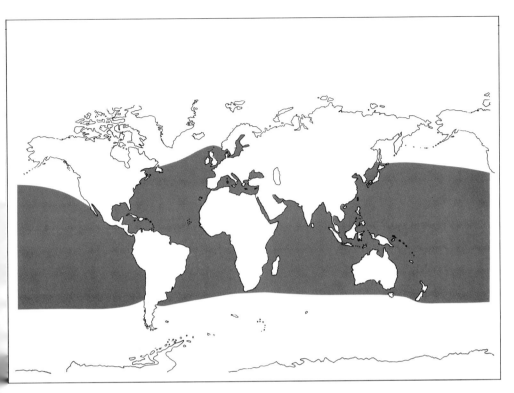

87. Distribution of Risso's dolphin, *Grampus griseus*

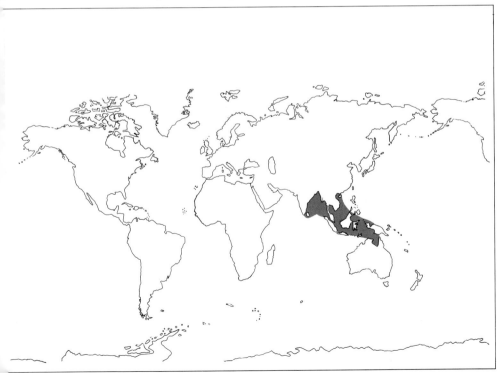

8. Distribution of the Irrawaddy dolphin, *Orcella brevirostris*

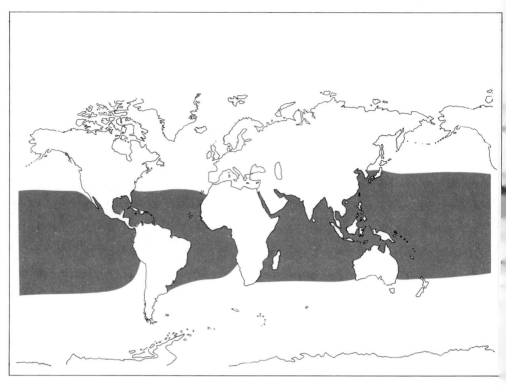

89. Distribution of Fraser's dolphin, *Lagenodelphis hosei*

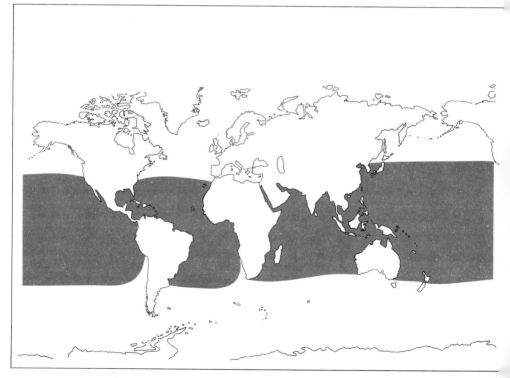

90. Distribution of the rough-toothed dolphin, *Steno bradensis*

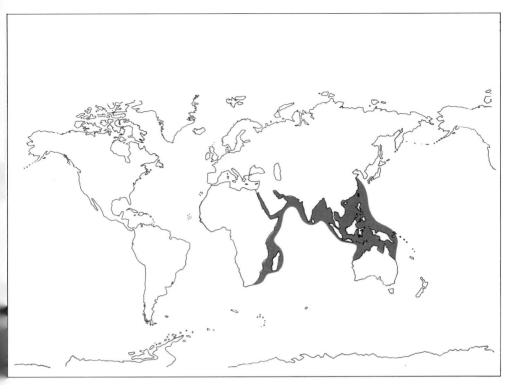

91. Distribution of the Indo-Pacific humpback dolphin, *Sousa chinensis*

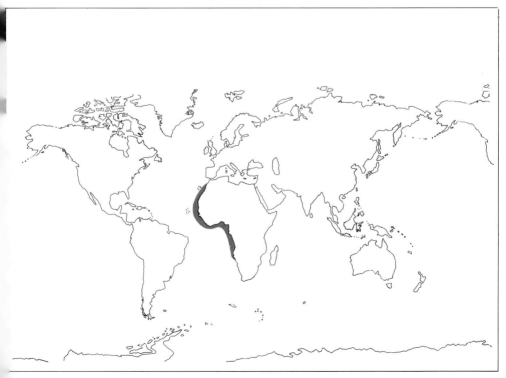

92. Distribution of the Atlantic humpback dolphin, *Sousa teuszii*

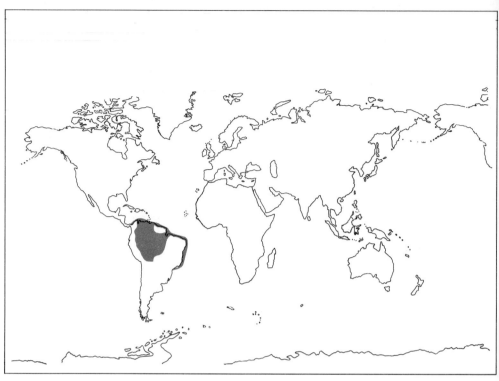

93. Distribution of the tucuxi dolphin, *Sotalia fluviatilis*

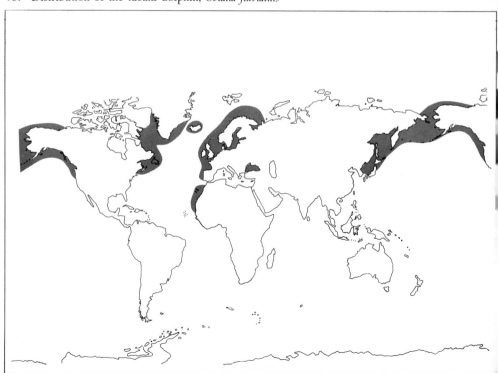

94. Distribution of the common porpoise, *Phocoena phocoena*

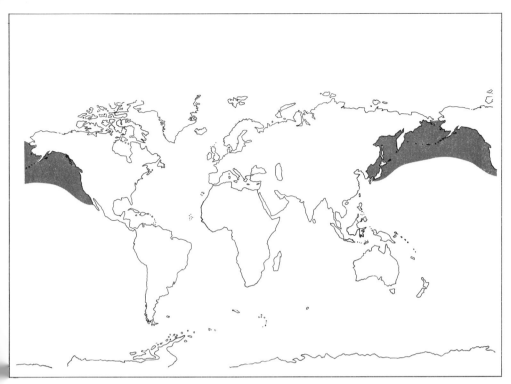

95. Distribution of Dall's porpoise, *Phocoenoides dalli*

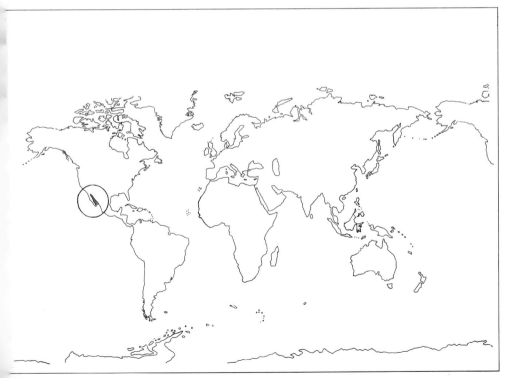

96. Distribution of the cochito porpoise *Phocoena sinus*

213

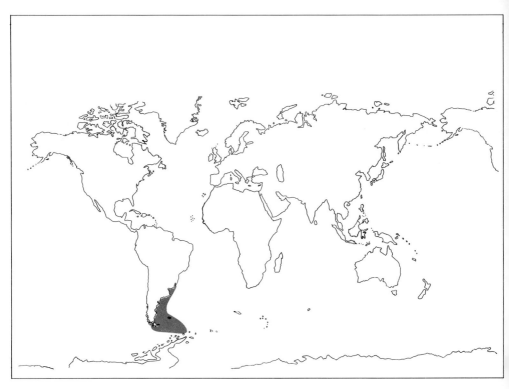

97. Distribution of the spectacled porpoise *Phocoena dioptrica*

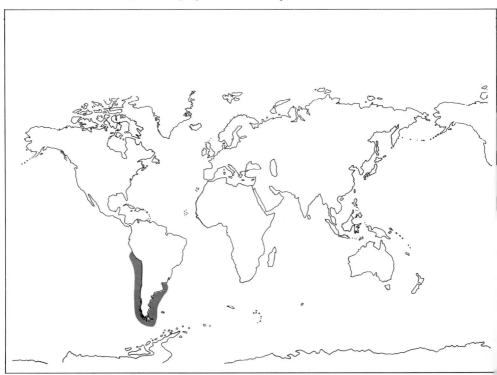

98. Distribution of Burmeister's porpoise, *Phocoena spinipinnis*

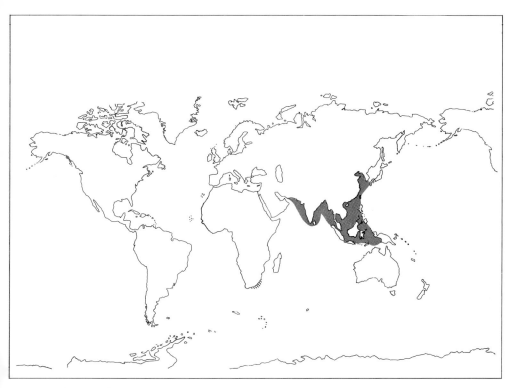

99. Distribution of the finless porpoise, *Neophocaena phocaenoides*

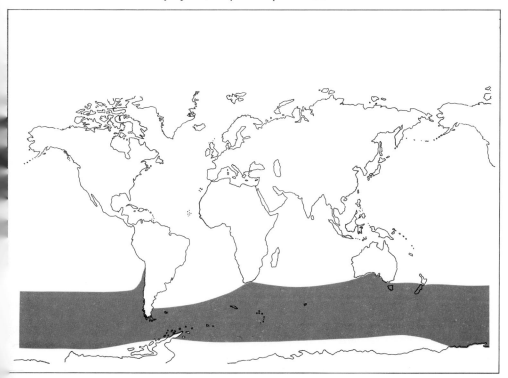

100. Distribution of Arnoux's whale, *Berardius arnuxi*

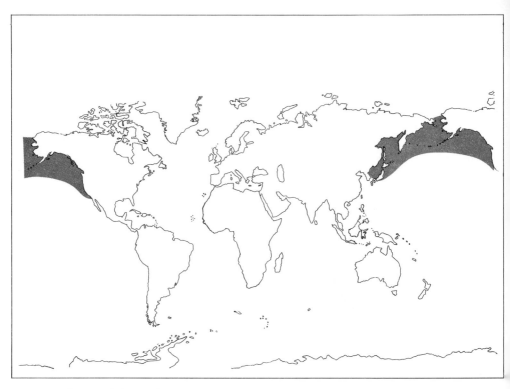

101. Distribution of the north Pacific giant bottlenose whale, *Berardius bairdi*

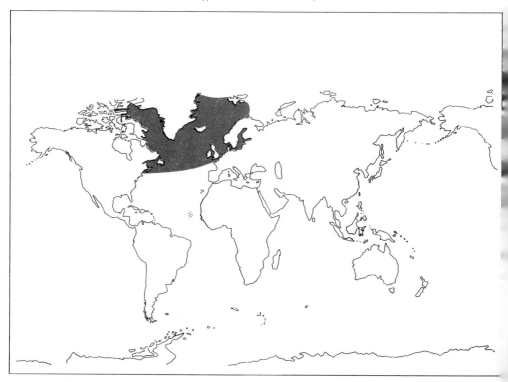

102. Distribution of the northern bottlenose whale, *Hyperodon ampullatus*

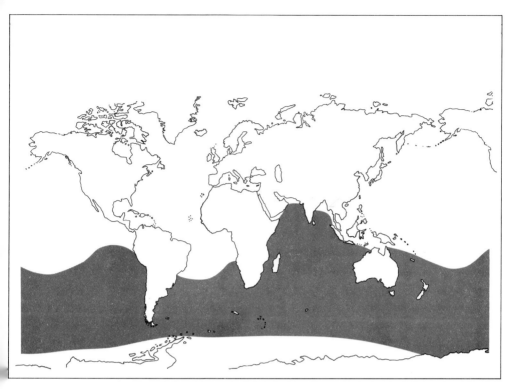

103. Distribution of the southern bottlenose whale, *Hyperodon planifrons*

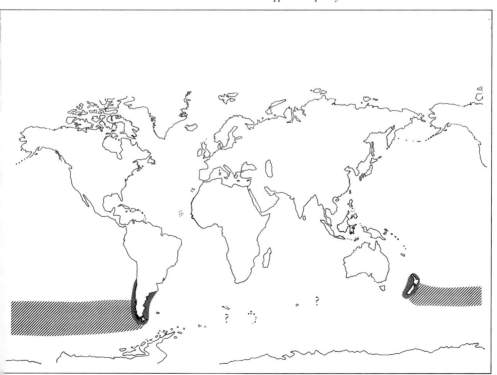

04. Distribution of Shepherd's beaked whale, *Tasmacetus shepherdi*

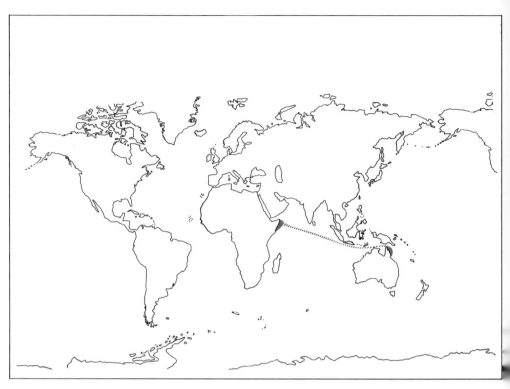

105. Distribution of Longman's beaked whale, *Indopacetus pacificus*

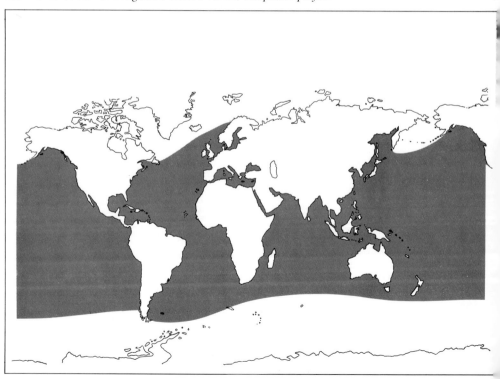

106. Distribution of Cuvier's whale, *Ziphius cavirostris*

218

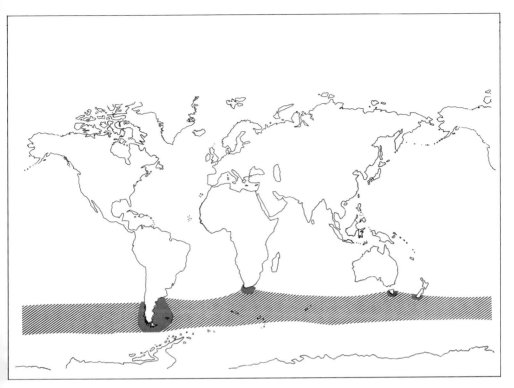

107. Distribution of Hector's whale, *Mesoplodon hectori*

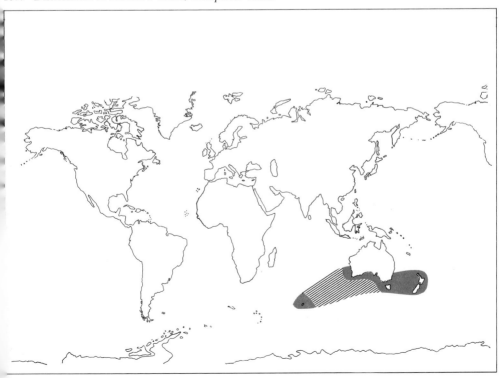

108. Distribution of the deep crest beaked whale, *Mesoplodon bowdoini*

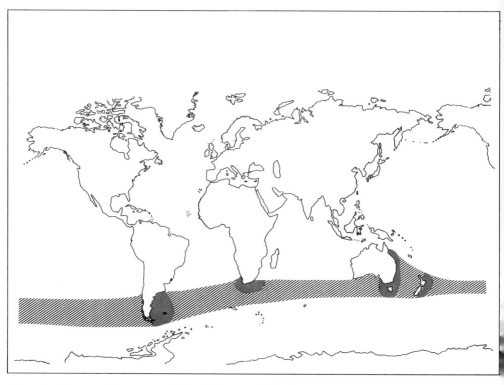

109. Distribution of the straptoothed whale, *Mesoplodon layardii*

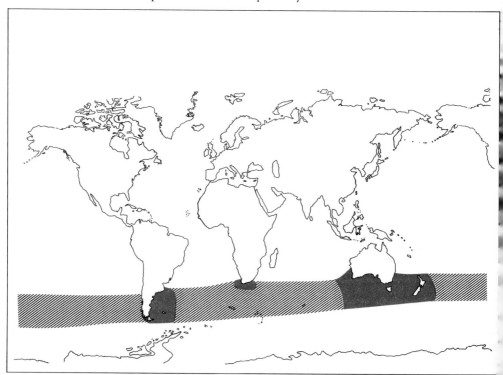

110. Distribution of the scamperdown whale, *Mesoplodon grayi*

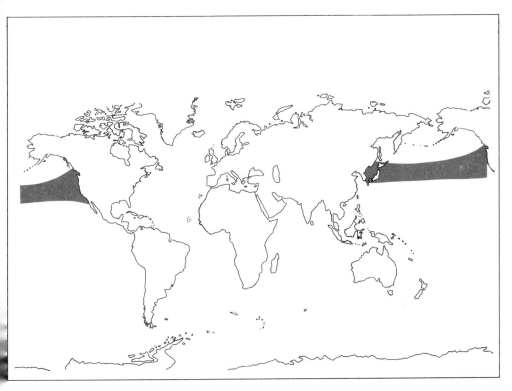

111.   Distribution of Hubb's beaked whale, *Mesoplodon carlhubbsi*

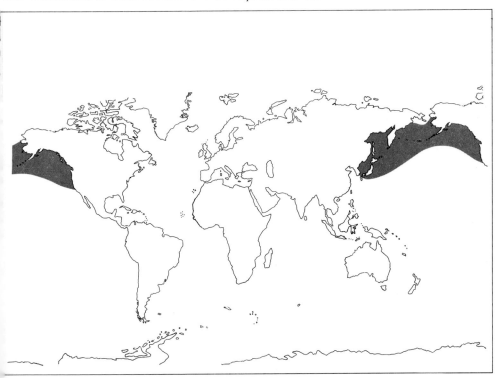

112.   Distribution of the sabretoothed whale, *Mesoplodon stejnegeri*

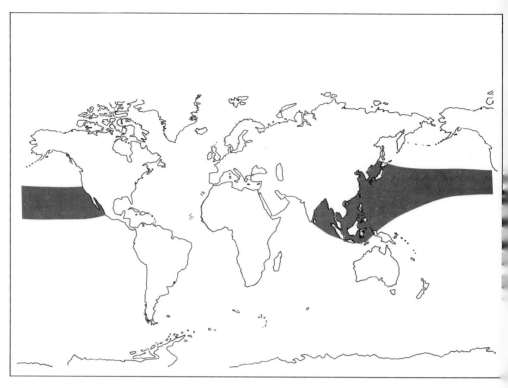

113. Distribution of the ginko-tooth whale, *Mesoplodon ginkodens*

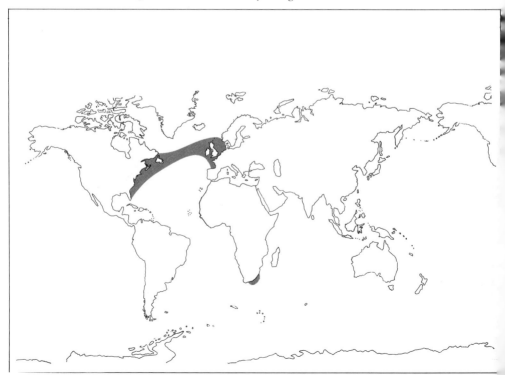

114. Distribution of True's beaked whale, *Mesoplodon mirus*

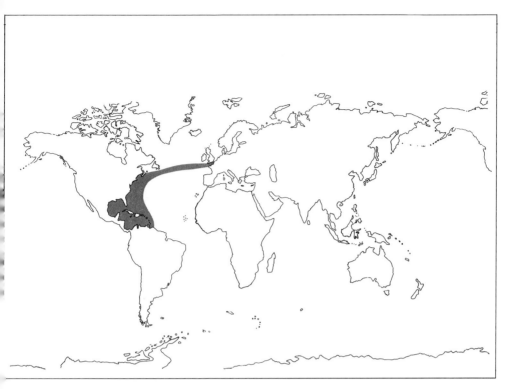

15. Distribution of the Antillean beaked whale, *Mesoplodon europeus*

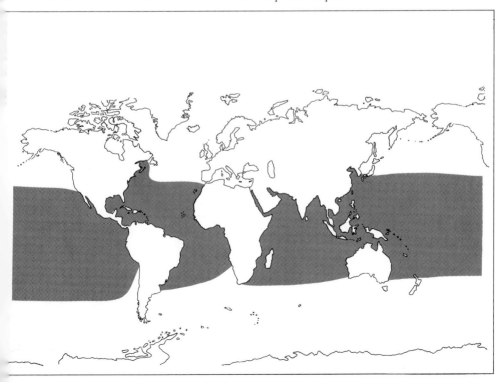

16. Distribution of the densebeak whale, *Mesoplodon densirostris*

223

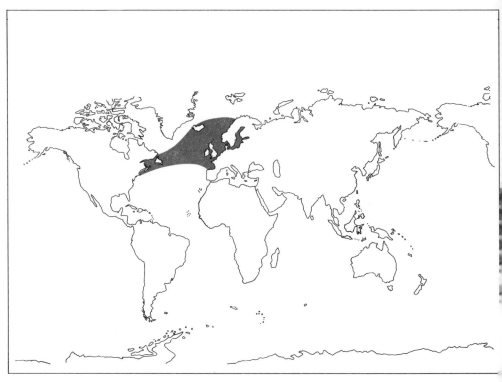

117. Distribution of the North Sea beaked whale, *Mesoplodon bidens*

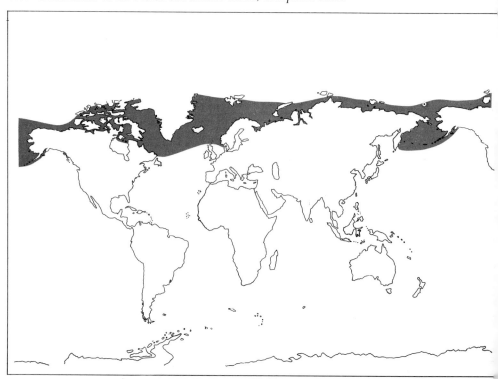

118. Distribution of the Narwhal, *Monodon monoceros*

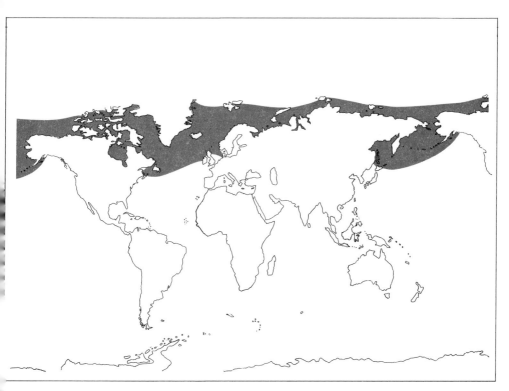

119. Distribution of the beluga or white whale, *Delphinapterus leucas*

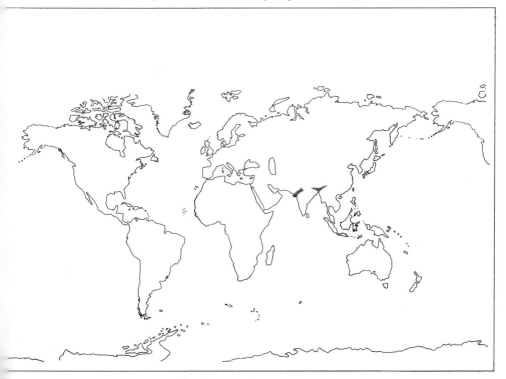

20. Distribution of the Ganges and Indus river dolphins, *Platanista gangetica*

225

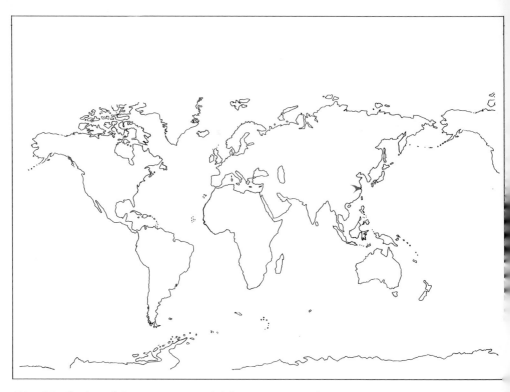

121. Distribution of the Yangtze river dolphin, *Lipotes vexillifer*

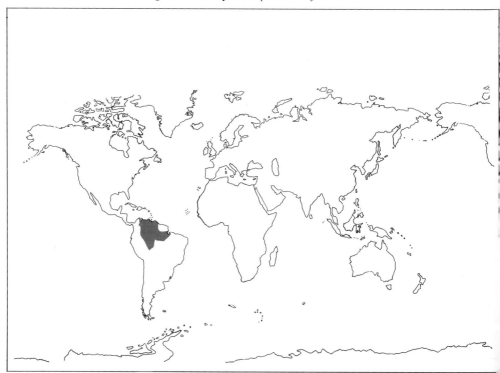

122. Distribution of the Amazon river dolphin, *Inia geoffrensis*

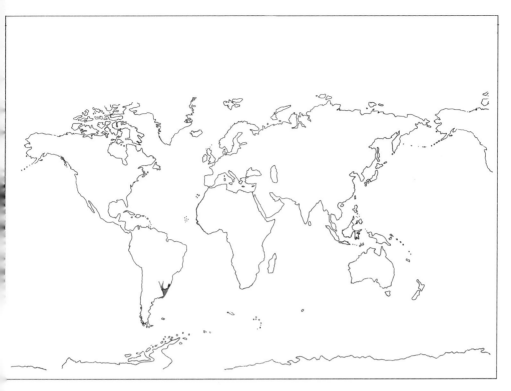

123. Distribution of the La Plata river dolphin, *Pontoporia blainvillei*

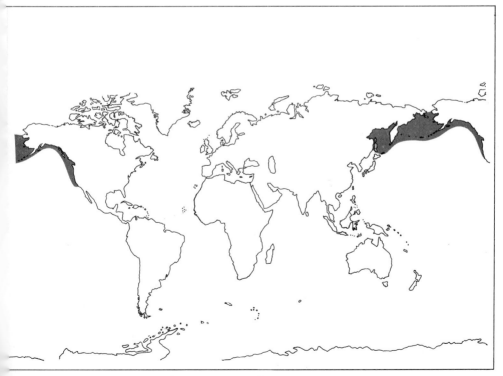

124. Distribution of Steller's sea lion, *Eumetopias jubatus*

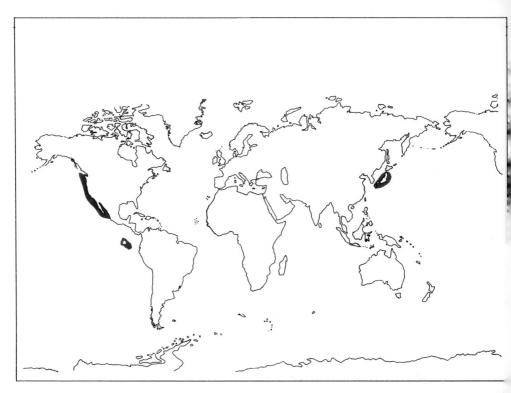

125. Distribution of the Californian sea lion, *Zalophus californianus*

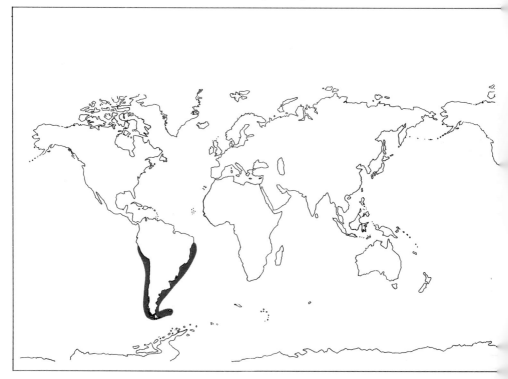

126. Distribution of the South American sea lion, *Otaria flavescens*

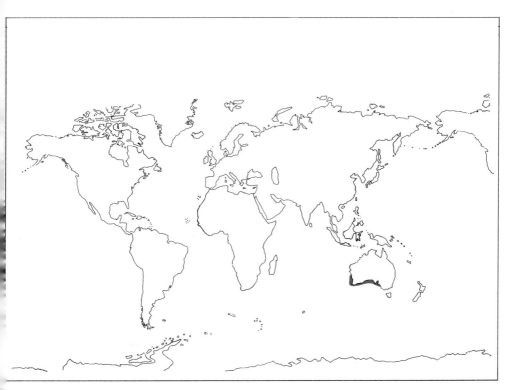

127. Distribution of the Australian sea lion, *Neophoca cinerea*

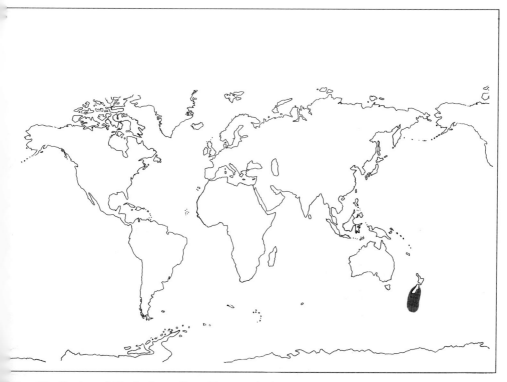

28. Distribution of Hooker's sea lion, *Phocarctos hookeri*

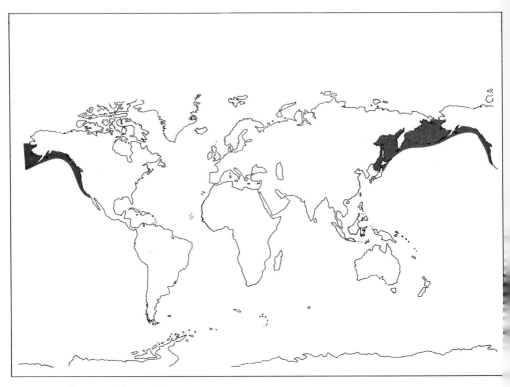

129. Distribution of the northern fur seal, *Callorhinus ursinus*

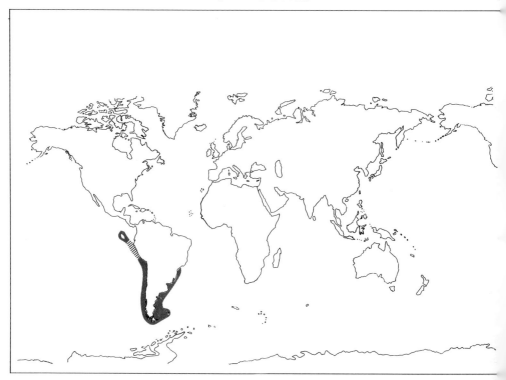

130. Distribution of the South American fur seal, *Arctocephalus australis*

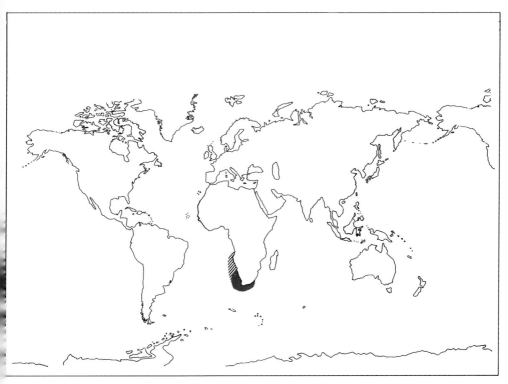

131. Distribution of the South African fur seal, *Arctocephalus pusillus pusillus*

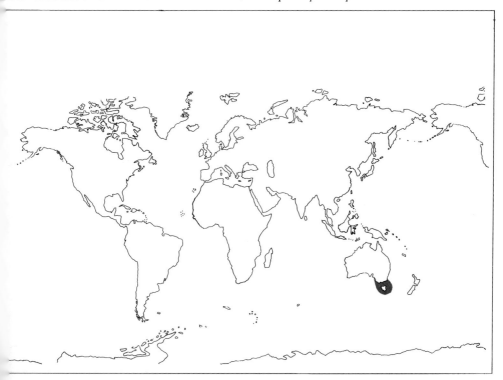

132. Distribution of the Australian fur seal, *Arctocephalus pusillus doriferus*

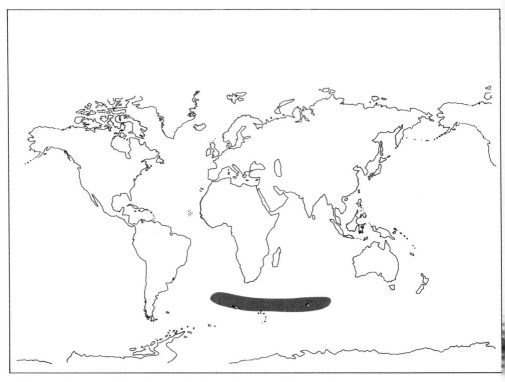

133.   Distribution of the subantarctic fur seal, *Arctocephalus tropicalis*

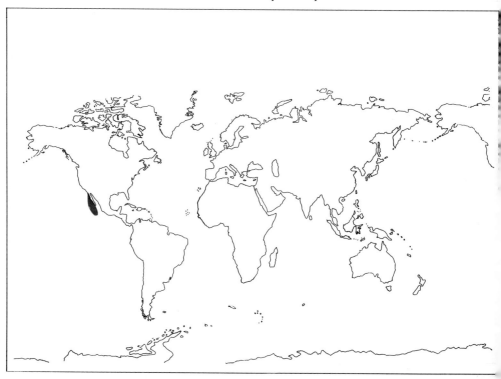

134.   Distribution of the Guadalupe fur seal, *Arctocephalus townsendi*

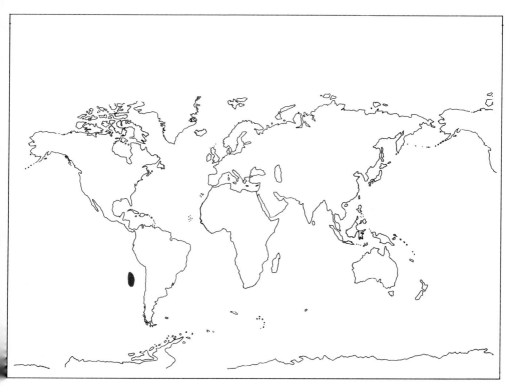

135. Distribution of the Juan Fernandez fur seal, *Arctocephalus philippi*

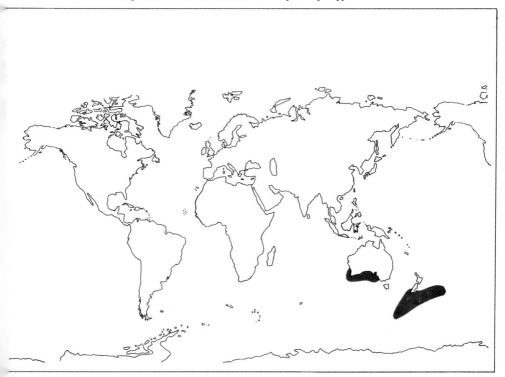

136. Distribution of the New Zealand fur seal, *Arctocephalus forsteri*

233

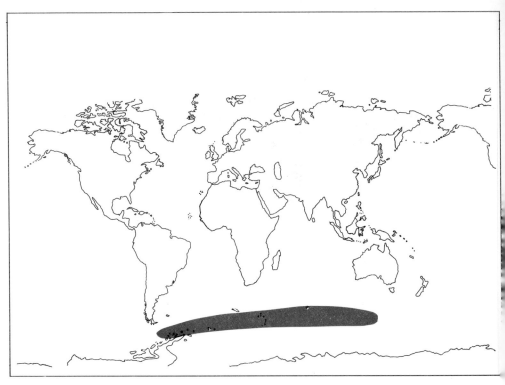

137.   Distribution of the Antarctic fur seal, *Arctocephalus gazella*

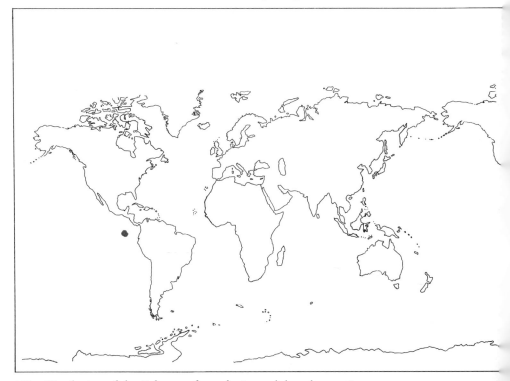

138.   Distribution of the Galapagos fur seal, *Arctocephalus galapagoensis*

234

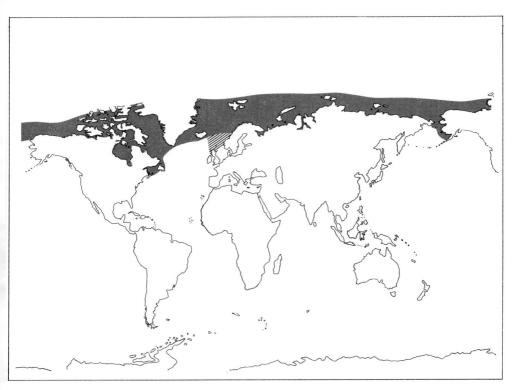

139. Distribution of the walrus, *Odobenus rosmarus*

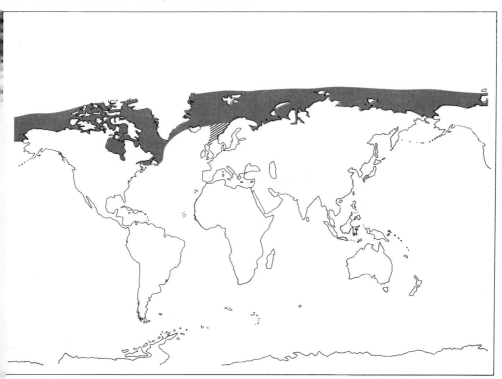

40. Distribution of the Bearded seal, *Erignathus barbatus*

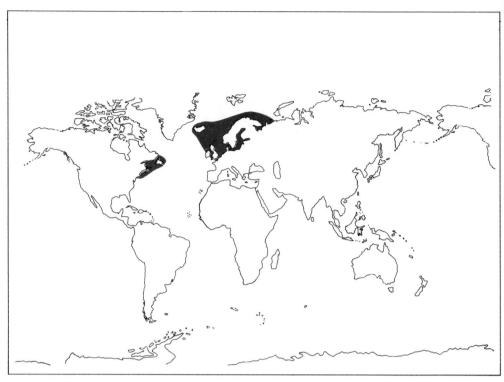

141.   Distribution of the grey seal, *Halichoerus grypus*

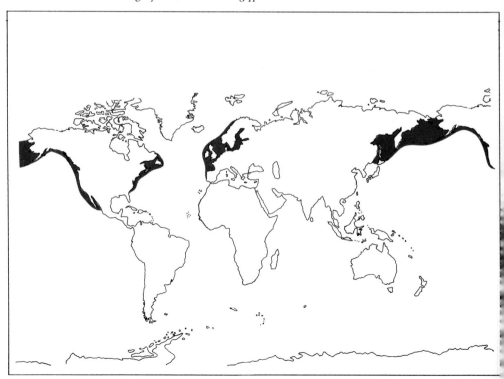

142.   Distribution of the common seal, *Phoca vitulina*

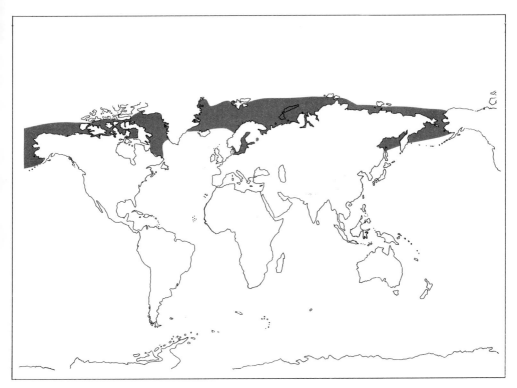

143. Distribution of the ringed seal, *Phoca hispida*

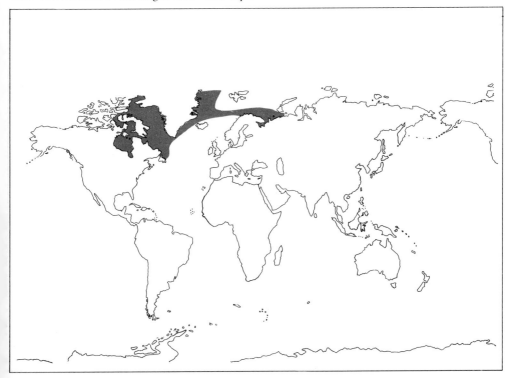

144. Distribution of the harp seal, *Phoca groenlandica*

237

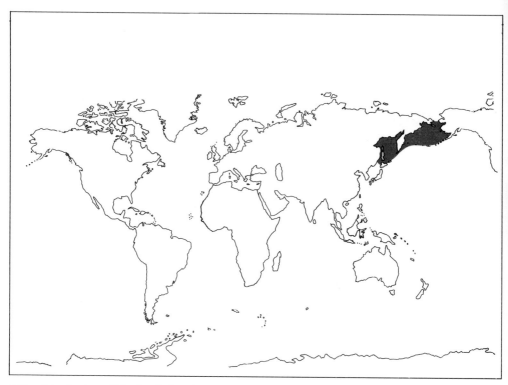

145. Distribution of the banded seal, *Phoca fasciata*

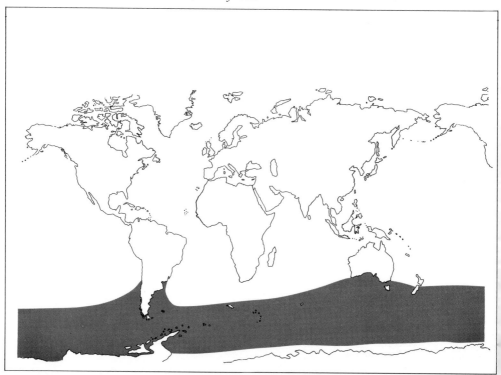

146. Distribution of the Weddell seal, *Leptonychotes weddelli*

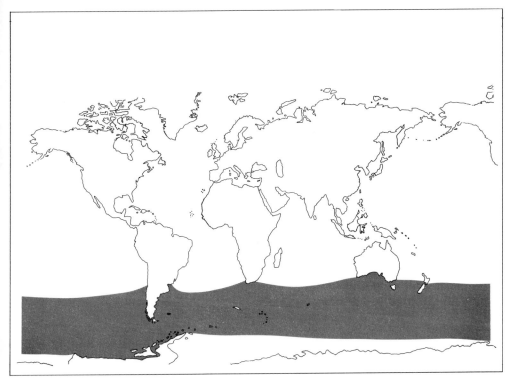

147. Distribution of the crabeater seal, *Lobodon carcinophagus*

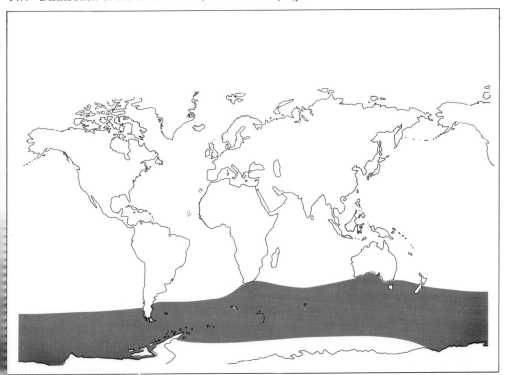

148. Distribution of the leopard seal, *Hydrurga leptonyx*

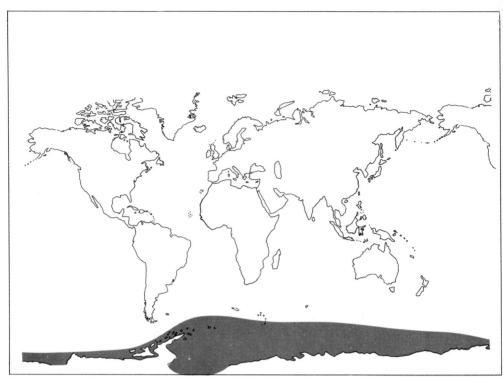

149. Distribution of the Ross seal, *Ommatophoca rossi*

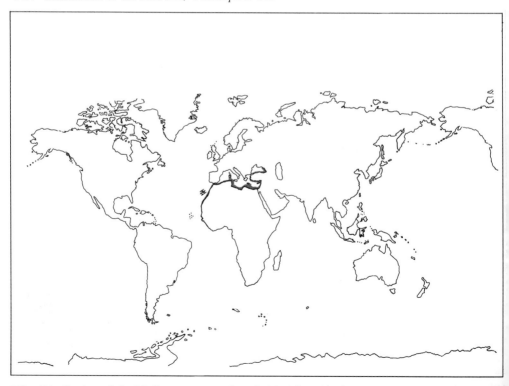

150. Distribution of the Mediterranean monk seal, *Monachus monachus*

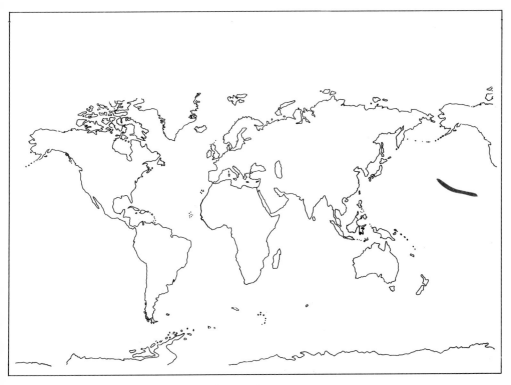

151. Distribution of the Laysan monk seal, *Monachus schauinslandi*

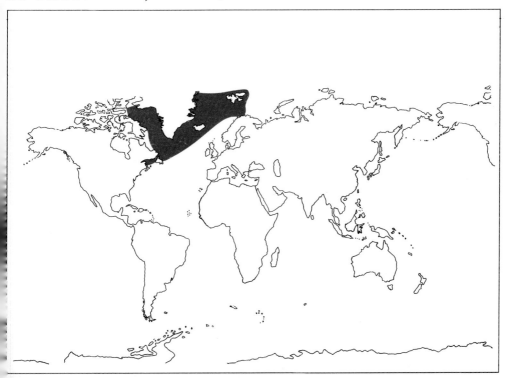

152. Distribution of the hooded seal, *Cystophora cristata*

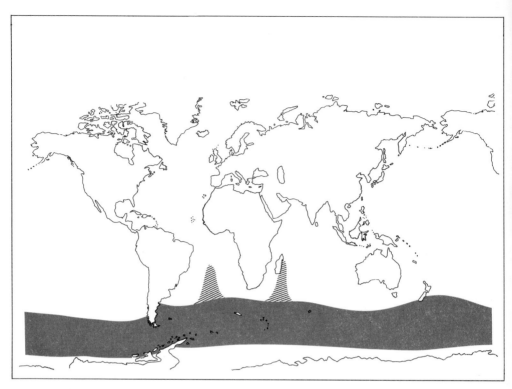

153. Distribution of the southern elephant seal, *Mirounga leonina*

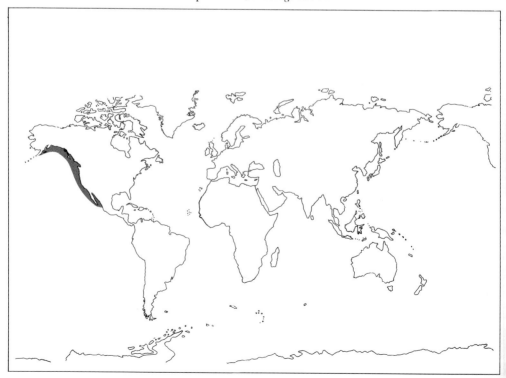

154. Distribution of the northern elephant seal, *Mirounga angustirostris*

242

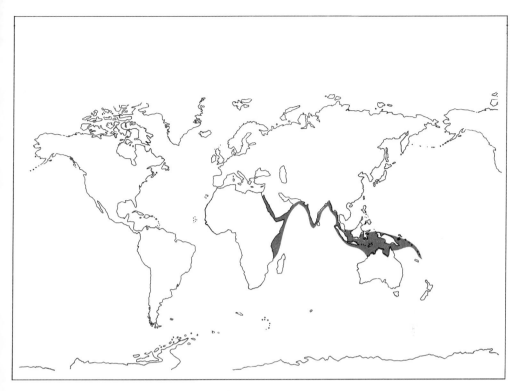

155. Distribution of the dugong, *Dugong dugong*

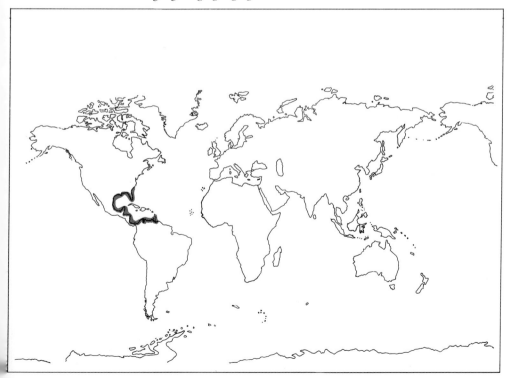

156. Distribution of the North American manatee, *Trichetus manatus*

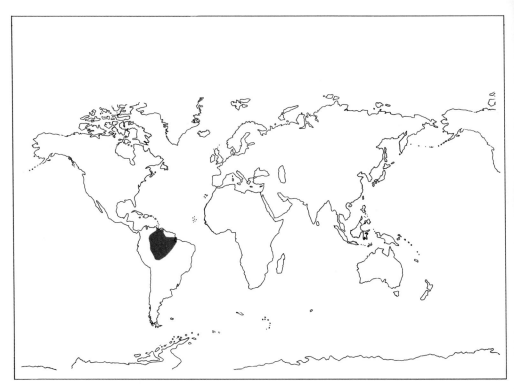

157. Distribution of the South American manatee, *Trichetus inunguis*

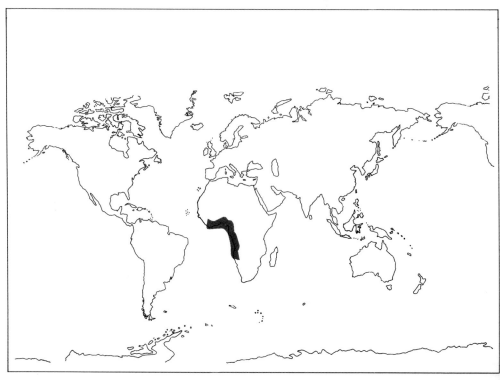

158. Distribution of the West African manatee, *Trichetus senegalensis*

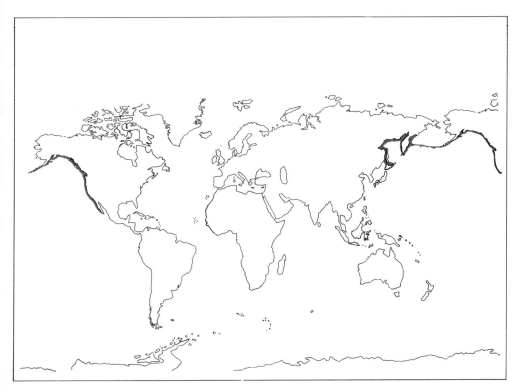

159.   Distribution of the sea otter, *Enhydra lutris*

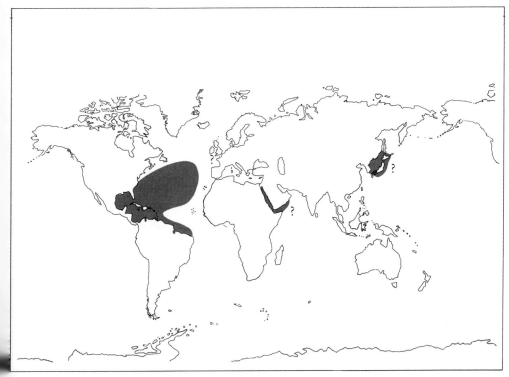

160.   Distribution of Gulfweed, *Sargassum* spp.

Appendix

The following is a list of books for those who wish to discover more about particular aspects of the marine world.

Barnes, R.D. *Invertebrate Zoology*, 4th ed. (Saunders, 1980)
This book is for the student or real enthusiast wishing to know much more detail about the invertebrates. All invertebrate groups are discussed from land, freshwater and the sea.

Bonner, W.N. *Whales* (Blandford Press, 1980)
This is a good general book about the whales. Bonner is also producing a series of books on the marine mammals.

Bustard, R. *Sea Turtles: Natural History and Conservation* (Collins, 1972)
Recommended for anyone wishing to know more about turtles.

Dunson, W.A. *The Biology of Sea Snakes* (University Press, 1975)
A detailed book mainly for the biologist.

Fincham, A.A. *Basic Marine Biology* (British Museum (Natural History)/Cambridge University Press, 1984)
A good general introduction to the subject although it requires some knowledge of basic biology.

Gaskin, D.E. *The Ecology of Whales and Dolphins* (Heinemann, 1982)
This is a good book for the enthusiast with some background knowledge and for the student.

George, D. and George, J. *Marine Life: An Illustrated Encyclopedia of Invertebrates in the Sea* (Harrap, 1979)
A good reference book for biologists and enthusiasts who do not mind scientific names.

Hardy, A. *The Open Sea: Its Natural History.* Part 1: *The World of Plankton*, 2nd ed. (Collins, 1970)
Although this is an old book it is worthwhile reading for enthusiasts, biologists and for general interest.

Herring, P.J. (ed.) *Bioluminescence in Action* (New York Academic Press, 1978)
This is a good book for the knowledgeable enthusiast and biologist.

King, J. *Seals of the World* (British Museum (Natural History))
This is for anyone wishing to know more about seals, sea lions and walrus.

Lythgoe, J. and Lythgoe, G. *Fishes of the Sea* (Anchor Press, 1975)
This gives information about the most common species found in the coastal waters of western Europe and the Mediterranean.

McClane, A.J. *McClane's Field Guide to Saltwater Fishes of North America* (Holt, Rinehart and Winston, 1965)
This book is sufficient for identifying some of the fish caught around the North American coast.

Marshall, N.B. *Developments in Deep-Sea Biology* (Blandford Press, 1979)
This is not written for the amateur but it is a valuable book for students and knowledgeable enthusiasts.

Menard, H.W. *Ocean Science: Readings from Scientific American* (W.H. Freeman and Company, 1977)
This collection of readings from *Scientific American* magazine deals with most oceanographic and marine biological subjects. They make fascinating reading for students and interested amateurs.
Newell, G.E. and Newell, R.C. *Marine Plankton: A Practical Guide*, 5th ed. (Hutchinson, 1972)
This book is for the student and enthusiastic plankton collector around the British Isles.
Watson, L. *Sea Guide to Whales of the World* (Dutton, 1981)
This is a good general guide to whales and dolphins though some of the common names have been changed from accepted forms.
Wheeler, A. *Fishes of the British Isles and North West Europe* (British Museum (Natural History), 1965)
A good guide to marine and freshwater fish around Europe.

Fish guide books are usually better bought in the local areas and many countries have guide books to the local marine life.

For further information about whales, dolphins and porpoises contact:

Mr. Denis McBrearty,
International Dolphin Watch,
Department of Anatomy,
University of Cambridge,
Downing Street,
Cambridge CB2 3DY
United Kingdom

The Marine Society, 202 Lambeth Road, London SE1 7JW, United Kingdom is always available to answer enquiries from seafarers throughout the world.

Scientists at the Natural History Museum in London will identify some specimens and answer enquiries; contact:

The British Museum (Natural History),
Cromwell Road,
London SW7 5BD
United Kingdom

Many British vessels carry meteorological logbooks; specimens and enquiries should be sent with the logbook or independently to:

The Meteorological Office Marine Division,
Eastern Road,
Bracknell,
Berkshire
United Kingdom

Glossary

Acari These are mites which are small (up to two millimetres) crawling animals with round bodies and four pairs of legs.

Alga A plant that reproduces using spores. The seaweeds are algae.

Ambergris A substance produced during digestion by the sperm whale. It is used commercially as a fixative for perfumes and make-up.

Amphibians A group of vertebrate animals living partly in water and partly on land; for example, frogs, toads and newts.

Annelids The name given to the group containing worms. The animals have rings on their bodies.

Antarctic convergence The region around the southern oceans at about 55 to 60 degrees south where cold low salinity Antarctic surface water flowing northwards meets warm tropical water and sinks beneath it.

Antenna The feelers, usually long, at the front of an animal; especially crustacenas. They are used as sensory detectors.

Anthozoa A group, belonging to the coelenterates, that includes the sea anemones and corals.

Baccy-juice water A yellow colour and bitter taste imparted to the water by a concentration of a certain species of phytoplankton. The term is used by fishermen; they know by the appearance of this water that there will be no herring to be caught in the area. (*See* **Red tide**)

Bacteria Minute organisms found everywhere. They are too small to be seen even with a simple microscope. Luminescent bacteria are common at sea.

Baleen The whalebone of mysticete whales. It acts as a filter to sieve food from the sea.

Bedlammers A corruption of 'Bête de la mer', meaning beast of the sea; it is a sealers' term for immature harp seals.

Bioluminescence (Sometimes called phosphorescence) The light created by certain animals and plants for courtship displays, warnings, and alarms.

Biosphere Refers to the entire living world: bio = living.

Bivalve The name of a group of molluscs having a shell in two halves, joined by a hinge.

Bloom (referring to a plankton bloom) The increase in growth of plankton when conditions are right. (*See* **Red tide**)

Breaching (usually with reference to whales and dolphins) The behaviour of leaping out of the sea and crashing back, probably in an effort to remove parasites and barnacles from the skin.

Brooding The protection of eggs by the adult. Sometimes a special pouch has evolved or protective flaps on the limbs or body.

Bryozoans A group of animals, commonly called sea mats, that grow flat on a surface or sometimes upwards like a plant. Seen close up the colony is made up of small boxes, each containing an individual.

Calcium carbonate A substance used to make up shells: commonly called limestone. Shell deposits on the sea floor make calcareous ooze.

Callosities Whitish horny growths found on the black or Biscayan right whale.

Carapace The cape-like cover over the front of some species of crustacean.

Cephalochordates A group in the chordates, which are animals with a backbone. Commonly called lancelets, they are eel-like and transparent.

Cephalopods A group of molluscs that includes the octopus, squid, cuttlefish and *Nautilus*.

Cetaceans The order to which all whales, dolphins and porpoises belong.

Chaetognaths A group of planktonic animals commonly called arrow-worms. They are about 2 centimetres long, with feathery fins at the sides and tail, and almost transparent.

Chlorophyll The green plant pigment that enables plants to carry out photosynthesis. The presence of chlorophyll labels an organism as a plant.

Chondrichthyes The group that covers the sharks, skates and rays. They are all cartilaginous fish.

Chordates The group that contains all animals with some kind of backbone or stiffening in the body at some time during their lives.

Cilia Stiff hairs that can beat to help carry a small animal along, or carry food or other material along a tube in a larger organism.

Cirripedia The group name for the barnacles.

Class A division, in the classification of organisms, below the Phylum.

Coelenterates The group of animals that includes jellyfish and sea anemones.

Colony A group of individuals working together, each doing different jobs so that in some cases there may appear to be only one large animal; for example, the Portuguese man-of-war, *Physalia*.

Crustaceans The group of animals with a hard outer skeleton that includes shrimps, crabs, barnacles and lobsters.

Ctenophores (Sea gooseberries) Small jelly-like organisms with eight rows of cilia down the sides.

Cubozoans A group of jellyfish, commonly called box jellyfish, that are found in the Indo-Pacific and are notorious for their extremely powerful and sometimes fatal sting.

Cyclostomes A type of primitive fish that has no jaw but a round mouth with teeth. They attach to fish and suck body fluid from their hosts. Cyclostomes are the lampreys and hagfish.

Diatoms A group of planktonic plants that live in cases made from silica. The cases may sometimes have spines, and are often found in chains.

Doliolids Transparent, barrel-shaped organisms belonging to the urochordate or sea squirt group of animals.

Dorsal The back surface of an organism.

Downwelling Occurs where water currents meet and sink beneath the surface.

Echinoderms The group name for starfish, sea urchins, brittlestars and sea cucumbers.

Echo-location The name given to a system of detecting the environment by using sound; as used, for example, by bats and whales.

Ecology The study of the animals and plants in their environment, their effect on each other, and the effect of their surroundings.

Exoskeleton An outer skeleton that is found in the crustaceans, insects and spiders.

Family A division in classification below Order.

Fertilisation The fusion of the male and female sex cells at the conception of a new individual.

Flagellates A group of minute organisms that move by the use of a whip-like tail called the flagellum.

Flipper The front limb of a whale or dolphin.

Fluke The rear limb of a whale or dolphin.

Formalin Fomaldehyde solution, used as a preservative.

Gastropods A group of molluscs that includes the snails and slugs.

Genus A division of classification below Family.

Gestation The length of time between the conception and birth of an organism. In humans it is the pregnancy period.

Gills The organs through which aquatic organisms obtain oxygen from the water.

Gribble The common name of a small crustacean wood-borer.

Habitat The place where an organism is living.

Hermaphrodite An organism that possesses both male and female reproductive organs.

Heteropods A group of swimming snails that swim upside down with the 'foot' acting as a fin.

Hydrozoans A group of coelenterates, usually living attached to a surface (*see* **white weed**) but there are notable exceptions: the Portuguese man-of-war, *Physalia* and the By-the-wind sailor, *Velella*.

Krill The old whalers' term for a shrimp-like animal that provides the major food for some baleen or whalebone whales.

Lancelet *See* **Cephalochordates**

Larvacea A group of sea squirts in which the young stage has developed the reproductive structures and there is no longer the typical adult sea squirt.

Lateral Refers to the side of an organism.

Lateral line The line of sense organs along the side of a fish that detects water movements and hence the presence of other organisms.

Malacostracans The class of crustaceans that includes the crabs, shrimps and lobsters.

Mantle The skin covering the body of a mollusc and lining the inside of the shell. The mantle is the outer covering of those that have no shell; for example octopus and squid.

Medusa The jellyfish stage of the coelenterate group.

Megalopa The advanced young stage of some crabs and shrimps.

Melon The front of the head of a dolphin or porpoise.

Mermaid's purse The common name for the egg-sac of some sharks, skates and rays.

Micron (μ) A measure of length for microscopic animals. It is one thousandth of a millimetre.

Mollusc The group of animals that includes snails, mussels, octopus and squid.

Mysticetes The group of great whales that have baleen or whalebone instead of teeth. They feed by filtering the food from the water through the baleen.

Nauplius The young stage of some species of crustacean.

Neap The name given to a small range of tides during the month that occurs between the spring tides.

Ocean skater The only oceanic insect. It is related to the freshwater pond skater often seen skipping across pond surfaces balancing delicately on the surface film.

Odontocetes The group of whales that have teeth.

Order A division of classification below Class.

Osteichthyes The group name of all bony fish.

Paralytic shellfish poisoning Caused by eating shellfish that have accumulated toxins from certain species of plant plankton that can cause paralysis and death in warm-blooded animals.

Phosphorescence *See* **Bioluminescence**.

Photophores The light-producing organs found on some species.

Photosynthesis The process that plants use to make food, using the energy of sunlight and carbon dioxide and water.

Phylum A major group in classification, below Kingdom.

Phytoplankton The plant plankton.

Pinnipedia The group name of the seals, sea lions and walrus.

Plankton The name given to organisms that drift with the currents.

Platyhelmintes A group of animals commonly called flatworms. They are usually small but visible to the naked eye.

Polyp Refers to a particular stage in the life-cycle of coelenterates, and to the living and feeding parts of coral and sea anemones.

Protozoans The single-celled animals.

Pteropods (Sea butterflies) Swimming snails in which the 'foot' has been modified into fins that resemble wings.

Pycnogonids Sea spiders.

Red tide Occurs where plankton have increased to such numbers that they change the colour of the water.

Rorquals A group of baleen whales that have a dorsal fin and are fast swimmers.

Salinity A measure of the amount of salt in water.

Salps Barrel-shaped animals that are the swimming sea squirts.

Sargasso Sea The area in the central North Atlantic in which the

Sargassum weed is found.

Scaphopods Molluscs that have a shell in the shape of an elephant's tusk, called Elephant tusk shells.

Scyphozoans The group of coelenterates that covers the true jellyfish.

Sessile Refers to an organism that is permanently attached.

Shipworm The common name of a mollusc that burrows into wood. These tunnels are much larger than those caused by the gribble, but the holes in the surface are smaller.

Silica The substance that some organisms use in building their cases. Deposits on the sea-bed form siliceous ooze.

Sirenians A group of mammals commonly called sea cows.

Species The division of classification below Genus, and the one that eventually describes one type of organism.

Spermaceti organ The organ in the head of the sperm whale, that is thought to assist during deep dives.

Spring The large range of tides occurring at full and new moon, that rise and fall between the highest and lowest water marks.

Spume The froth on seawater caused by rough weather. The bodies of planktonic organisms are smashed, and release the oil that forms the froth.

Spy-hopping The habit, of many whales and dolphins, of 'standing' with their heads out of the water, looking around.

Tidal Range The height, between the high and low water marks, that the tide covers during its rise and fall.

Trochophores The young stages of some species of annelid worm and mollusc.

Upwelling Where the water masses from deep down are brought to the surface. This is caused by submarine ridges, offshore surface winds, or two ocean currents diverging from each other.

Urochordates The group of chordates that are commonly called sea squirts. Usually only the young stages have the important simple backbone.

Veliger The advanced young stage of molluscs.

Ventral Refers to the underside of organisms.

Vertical migration The behaviour of many organisms that migrate up and down in the water, usually in a twenty-four hour cycle.

White weed The common name given to a growth of hydrozoans or sea firs.

Zoea The young stage of some species of crabs and shrimps.

Zooplankton The animal plankton.

Index

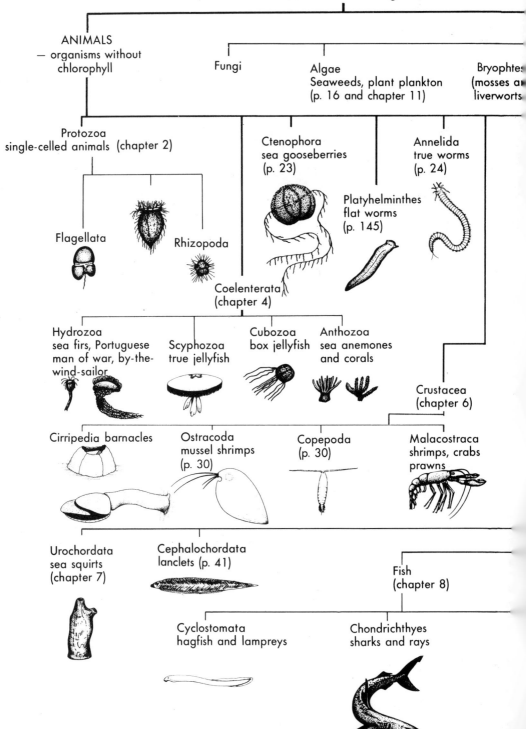

BIOSPHERE – the living world

ANIMALS
— organisms without
chlorophyll

Fungi

Algae
Seaweeds, plant plankton
(p. 16 and chapter 11)

Bryophtes
(mosses a
liverworts

Protozoa
single-celled animals (chapter 2)

Ctenophora
sea gooseberries
(p. 23)

Annelida
true worms
(p. 24)

Platyhelminthes
flat worms
(p. 145)

Flagellata

Rhizopoda

Coelenterata
(chapter 4)

Hydrozoa
sea firs, Portuguese
man of war, by-the-
wind-sailor

Scyphozoa
true jellyfish

Cubozoa
box jellyfish

Anthozoa
sea anemones
and corals

Crustacea
(chapter 6)

Cirripedia barnacles

Ostracoda
mussel shrimps
(p. 30)

Copepoda
(p. 30)

Malacostraca
shrimps, crabs
prawns

Urochordata
sea squirts
(chapter 7)

Cephalochordata
lanclets (p. 41)

Fish
(chapter 8)

Cyclostomata
hagfish and lampreys

Chondrichthyes
sharks and rays